SpringerBriefs in Applied Sciences and Technology

T0075367

Computational Intelligence

Series Editor

Janusz Kacprzyk, Systems Research Institute, Polish Academy of Sciences, Warsaw, Poland

SpringerBriefs in Computational Intelligence are a series of slim high-quality publications encompassing the entire spectrum of Computational Intelligence. Featuring compact volumes of 50 to 125 pages (approximately 20,000–45,000 words), Briefs are shorter than a conventional book but longer than a journal article. Thus Briefs serve as timely, concise tools for students, researchers, and professionals.

More information about this subseries at https://link.springer.com/bookseries/10618

Daniel Albiero

Robots and AI: Illusions and Social Dilemmas

Cartesian Illusions and Social Unsettling
Questions

 Springer

Daniel Albiero ⓘD
School of Agricultural Engineering
University of Campinas—UNICAMP
Campinas, Brazil

ISSN 2191-530X   ISSN 2191-5318 (electronic)
SpringerBriefs in Applied Sciences and Technology
ISSN 2625-3704   ISSN 2625-3712 (electronic)
SpringerBriefs in Computational Intelligence
ISBN 978-3-030-95789-6   ISBN 978-3-030-95790-2 (eBook)
https://doi.org/10.1007/978-3-030-95790-2

This Springer imprint is published by the registered company Springer Nature Switzerland AG
The registered company address is: Gewerbestrasse 11, 6330 Cham, Switzerland

I dedicate this book to my daughter Isis, may she master AI and not the other way around.

Foreword

An agricultural engineer for 20 years, Daniel Albiero, after completing his doctorate at the State University of Campinas (Unicamp), chose to pursue an academic career, dedicating himself to teaching and research. This choice, it was already possible to predict, would be successful. The restlessness and questioning attitude demonstrated in classrooms and laboratories during undergraduate and graduate studies at the Faculty of Agricultural Engineering—Unicamp indicated a critical, reflective, and innovative profile, essential characteristics for professionals who set out to generate knowledge, always seeking to improve people's quality of life.

After completing his doctorate, the first opportunity arose to follow the university paths of teaching at the Federal University of Ceará, a prosperous state in Northeastern Brazil, where Daniel would remain for about eight years. Until then, his whole life had been spent in the state of São Paulo, the most developed in Brazil. Indeed, a significant contrast would decisively polish the teacher's education, convictions, principles, values, and concepts.

A notice from Unicamp brought him back to Campinas in 2018 to take over as a professor at the Faculty of Agricultural Engineering, where he had completed his professional and academic training.

Since the undergraduate research project, the chosen area was an agricultural machinery, which remained in postgraduate studies and led to admission as a professor at the university. In the Ceará context, the work was directed toward agroecological machines, alternative energies, family farming, and other issues related to technological development linked to social and human issues due to the conditions and needs of the region.

It did not take long to conclude that the accelerated technological development provided by robotics and artificial intelligence would be essential for the future of their areas of activity in agricultural engineering. Accordingly, he dedicated himself to getting to know the technical foundations that underpinned the advancement of knowledge in this field. As this occurred, a vast spectrum of possibilities emerged and, above all, as solutions to the problems existing in the Brazilian agricultural system.

He soon realized the need to properly equate the impacts resulting from the practical application of robotics and artificial intelligence in social issues. As a university professor, he began to reflect on the importance of comprehensive training for professionals who would dedicate themselves to the subject, incorporating humanist concepts essential to understanding the system as a whole.

This brief background helps you understand the motivations that defined and shaped this book, its approaches, emphases, and goals. Furthermore, they allow considering the context of reflections on technological and philosophical themes, which are related and intertwined at all times, showing that they are inseparable and decisive for a better future, adequate quality of life, and social justice.

From the opening pages, by situating robotics and artificial intelligence as fruits of science, Daniel Albiero leads us to critically reflect on interactions with social issues and the logic of interactions with human values and consequences. While presenting fundamental concepts, he often questions them, presenting convincing arguments to demonstrate that technology has its limits established by biological diversity and the moral principles of human beings.

The author navigates through considerations about the scientific method, flies over the nature of social phenomena, and immerses himself in chaos theory, always with thought-provoking, provocative, and questioning observations. It involves the reader in a universe full of intersections and possible paths to be explored by the imagination, regardless of the level of knowledge acquired on the topics discussed.

With great skill, Daniel analyzes the impacts of robotics and artificial intelligence on society. Moreover, vice versa, as he proposes, characterizing this relationship as two way. Finally, presenting a coherent line of reasoning, it goes through some traditional professions—doctor, lawyer, and teacher—to demonstrate the many changes that could occur, without them implying the extinction of these professionals, who will continue to be necessary because of the existing human relations in the preservation of life, justice, and learning.

Ethical issues could not be overlooked in this work. More than seventy years ago, Isaac Asimov established in his works of fiction (?) the three laws of robotics, based on human ethical principles and guaranteed harmony and the limits of coexistence. Without going into the field of science fiction, Daniel discusses the subject, relating it to democracy and human rights, indicating that ethics alone will not be enough to avoid dystopias that could compromise our future.

Throughout the book, there is a clear impression that Daniel Albiero wants to motivate the reader to learn more about robotics and artificial intelligence. However, it gets more than that! For those related to the theme, reflections are precious to consolidate certainties and deepen doubts, for those who know little, an opportunity to start the learning process correctly.

The patron of Brazilian education, Paulo Freire, known in many countries for the pedagogical methods he developed and applied, is referenced in this work to analyze the importance of robotics and artificial intelligence in education and emphasize that the teacher will continue to be indispensable. He should keep in mind, during his reading, one of his historical phrases: *"There is no knowing more or knowing less: there is different knowledge."*

Campinas, Brazil

José Tadeu Jorge
Dean of the School of Agricultural
Engineering—FEAGRI (1987–1991 e 1999–2002)
Rector of the University of Campinas—UNICAMP
(2005–2009 e 2013–2017)

Preface

The origin of this book is interesting. It all started with an invitation to write a chapter on robotics initiation for kindergarten and elementary school teachers in a content book sponsored by the National Rural Learning Service of the State of Paraná-Brazil (SENAR-PR). The invitation was for a 20-page chapter. However, this author is very wordy, and 120 were delivered!

The editors of that book liked the proposal and the language and accepted the chapter, but the SENAR board considered inadequate parts…some too long…others too colloquial…others perhaps too ideological. So, cuts were requested. Of the 120 original pages, only 40 were published in SENAR's book, one of the cut parts (40 pages) was the origin of this work.

However, the path taken for this text has not yet been that of a book. After rejecting the initial manuscript for SENAR, the work was redesigned into a review article format and submitted to the International Journal of Social Robotics (SORO-Springer). The editor-in-chief found the material exciting and sent it to SORO's excellent ad hoc reviewers. They made many criticisms, some soft, some harsh, but all constructive. When starting the corrections and following the suggestions, it was noticed that the article would increase a lot in size, making publication in a scientific journal unfeasible.

So, I made a bold decision and contacted Dr. Thomas Ditzinger, Editorial Director Interdisciplinary and Applied Sciences, Engineering in Springer. Dr. Ditzinger liked the material and sent it to Prof. Dr. Janusz Kacprzyk Series Editor *Springer Briefs in Computational Intelligence*, who approved the proposal of this book. Thus, I made the corrections suggested by the ad hoc reviews of SORO, and from 40 pages, the manuscript increased to 70 pages.

The main idea is to frontally attack the Cartesian worldview (shortsighted and fragmentary) against the advent of robotics and AI. The focus is to present these cutting-edge technologies with a social and historical bias within the scope of the continuous development of humanity, as every era is capable of encompassing the technology in which it is ready to absorb, at the beginning of the twenty-first century, without a shadow of a doubt we are ready for robotics and AI. However, this state of readiness does not exempt us from evaluating the virtues and vices of these technologies, especially considering the human being as a whole, not fragmentary, and not only a privileged few but all social classes.

This time in Chap. 1, I make a basic introduction, listing the conceptual assumptions and locating the subject in the current and world context facing humanity's technological and social development since our beginnings.

In Chap. 2, I dive into the sea of chaos (mathematical theory) and demonstrate how inappropriate it is to treat robotics and AI in a Cartesian context, especially considering the properties of chaos with its self-similar emergent structures.

Chapter 3 is a walk through the transition zone far-from equilibrium (social and thermodynamic) of this era of extremism and extremists it tries to present the illogic of extremist positions and mainly lists the focus of balance.

The disruptions and social and technological changes arising from robotics and AI are outlined in Chap. 4, where I present passionate defenses of the three primary professions on the planet, not coincidentally, the professions supposedly most "threatened" by robotics and AI.

Moreover, finally, I conclude this book with a call to hope and love. Amazingly, I genuinely believe that only with Agape we can have a future with freedom within this increasingly robotic world.

Campinas, Brazil   Daniel Albiero
October 2021

Acknowledgments

I would like to thank SENAR-PR specifically the Agrinho Program for the invitation to write the chapter on robotics for kindergarten and elementary school teachers.

This work, initiated in 2018, was carried out within the Federal University of Ceará (UFC), the end of the writing was in 2021 in the University of Campinas (UNICAMP), and I thank UFC and UNICAMP for the support.

I thank my wife Danila and my daughter Isis for their patience in sharing me with robots and AIs.

I am grateful to the anonymous reviewers of the International Journal of Social Robotics (Springer) for valuable comments.

I thank the publishers of Cambridge University Press for permission to use Schrödinger's epigraph in Chap. 1 and EPBM for permission to use Mandelbrot's epigraph in Chap. 2.

I thank the funding agencies for the financial resources: CNPq: 304931/2015-2; FUNCAP: AE1-0079-000250100/13; CAPES: 2982/2013; FAPESP: 19/06078-8.

And last but not least, I thank the God of chaos and the God of religions. One leads nature, another leads humanity. In a non-Cartesian hypothesis, maybe they are the same entity.

Acknowledgments

Contents

Contents

About the Author

Prof. Dr. Daniel Albiero is the chair of machinery and robotics in agriculture in the School of Agricultural Engineering (FEAGRI) at the University of Campinas (UNICAMP). He holds a Ph.D. in Agricultural Engineering from FEAGRI/UNICAMP in 2009, a Master in Agricultural Machinery from FEAGRI/UNICAMP in 2006, and an Agricultural Engineer from FEAGRI / UNICAMP in 2001.

From 2002 to 2003, he was a technical assistance engineer at Agroamazônia/John Deere—Brazil. From 2003 to 2005, he was the engineering coordinator of Empresa Hibrema. From 2009 to 2010, he was the engineering supervisor at Company Schalke/EMME2/HIBREMA-Germany.

He was a professor of design of machinery and energy in agriculture at the Center for Agricultural Sciences (CCA) of the Federal University of Ceará (UFC) from 2010 to 2018, was the coordinator of the Graduate Program in Agricultural Engineering at UFC (level 5 at CAPES) in the period 2015–2017, was a representative of CCA/UFC on the technological innovation committee of the UFC's Rector of Research, and was a representative of CCA/UFC on the committee for the implementation of the UFC technology park by the UFC's Rector.

In July 2018, he assumed machinery and robotics projects in agriculture at FEAGRI/UNICAMP. Currently, he is the extension coordinator at FEAGRI/UNICAMP in the period 2019–2022. He has been a CNPq productivity scholar in Technological Development and Innovative Extension (DT-2) since 2013 with a current scholarship until 2021.

He published more than 110 peer-reviewed journal articles, has 29 patents filed, six patents, and two licensed patents. He has more than 370 works published in national and international congresses and is the author of four books. In addition, he coordinated the development, design, and construction of more than 70 machines, robots, and renewable energy systems prototypes.

He was the adviser for more than 100 undergraduate students, IC with scholarship, voluntary IC, CBT, monographs, specialization, PED, PAD, and tutoring. In addition, he completed the supervision of eight master's dissertations and five doctoral theses.

Professor Albiero won three INOVA-UNICAMP Inventors awards for granted patents. He was honored with praise by the Federal University of Ceará for the services provided. He is the coordinator of two international agreements between UNICAMP/UFC with German educational institutions: 1-International partnership with Wuppertal University (Germany) with Professor Benedikt Schmuelling and subjects: power electronics, electric mobility, energy storage, and automations in tractors; 2-International partnership with TH-Koln (Germany) with Professor Ramchandra Bhandari and subjects: renewable energy, alternative fuels, and alternative foods. It has a research and development partnership with eight companies.

Full curriculum in:
CV LATTES: https://orcid.org/0000-0001-6877-8618

Chapter 1
Introduction

> Science cannot tell us a word about why music delights us, of
> why and how an old song can move us to tears.
> *Erwin Schrödinger.*

Robotics and Artificial Intelligence (AI) are waves that reach the beach of society with increasingly intense transforming effects, such as "Ulmo, Valar, and Lord of Waters" when he showed himself to Tuor, son of Huor in front of Vinyamar [1]. Its discussion presents itself with all the contradictions that exist in an invitation that cannot be denied.

In this context, Robotics and AI elements inserted in contemporary society are based on science, backed by the scientific method and directed by development and progress of our civilization. Several authors state that, at the beginning of the twenty-first century, we are living in times of technological disruption [2–7]. So it is an obvious deduction that these disruptive technologies have an influence on the structure of our society, mainly in the formation of a worldview.

However, this scenario is not new, because ancient societies were no different from the current one concerning technical "revolutions," although they spent all their time on activities related to survival. The difference is that today we have more choices and more freedom of movement because each historical phase's technology has allowed the expansion of the human capacity to dominate nature [8]. New, maybe to be some terminological premises that must be defined as emphasizing society, robotics, and artificial intelligence, assumptions that can lead to a technopositivist worldview if they are based on cartesian illusions.

Will be used [9] definition of society, which considers it as any system of human interactions culturally standardized according to symbols, values, norms, positions, and roles occupied and used by the members of a group. The groups to which all subsequent discussion refers are modern industrialized societies where social institutions conceptualized by Durkheim precede the existence of any individual and exercise limitations on them [10]. However, in our definition, it will not be disregarded that society is not "external" to its members, because although it is external to each individual if seen in isolation, it cannot be external to all individuals collectively, that is, social facts restrict what we do, but do not determine what we do [10].

D. Albiero, *Robots and AI: Illusions and Social Dilemmas*,
SpringerBriefs in Computational Intelligence,
https://doi.org/10.1007/978-3-030-95790-2_1

The criterion used to define robotics is [11], who describe it as the science that studies the intelligent connection between perception and action. According to the book Handbook of Robotics edited by them, the operation of a robotic system is entrusted to a locomotion system to move around in the environment and a manipulation apparatus to operate objects where actuators animate the mechanical components of a robot. There to be a robotic action, a system must rely on perception, which is extracted from sensors that provide information on the state of the robot (position and speed) and of the environment (strength, range, vision), but also with the intelligent connection, which is provided by a digital processing system. In this digital processing system, computational programming is based on a control architecture that explores the learning and skills of information acquired through a defined logic.

Concerning Artificial Intelligence, according to [12], reasoning technology involves inference, planning, and learning, which are based on symbolic reasoning using First-Order Predicate Logic (FOPL) integrated with Bayesian Probability Theory. Bayesian logic deals with the representation of uncertainties about propositions, especially when there is no complete information about the plausibility of the proposition [13]. Thus, the main element resulting from the merger between robotic methods and AI is the so-called robot that can be defined as an automatic system that can make decisions [14]. References [15–17] define an ontology for autonomous robotic systems. The specification of the concept of an autonomous robot is explained through a logical theory where the robot can feel the environment and plan actions based on the environment to achieve specific goals without external control. These specific objectives are achieved by the operationalization of functional modules, described by Langley et al. [18]: Recognition and categorization; Decision-making and choice; Perception and situation assessment; Prediction and monitoring; Problem-solving and planning; Reasoning and belief maintenance; Execution and action; Interaction and communication; and Remembering, reflection, and learning. References [19, 20] claim that depending on the level of autonomy, these systems will have varying degrees of interaction with other intelligent agents.

The term technopositivism is defined by Roberson [21] as a naive faith in the "promises" of technology. According to Roberson [21]: *"Technopositivism taps into the human psyche, into our optimism and our desire to find external, mechanical solutions to complex, very human problems. The virtues of every emerging technology have been overestimated not only by those who stand to gain financially from their adoption but also by those who acquior wish to acquire them."*

Reference [22] in his excellent and demystifying book declares that this kind of "mythology" about AI is bad because it covers a scientific mystery with an endless talk of the ongoing process; this mythology proposes the belief that these myths are inevitable successes, but genuine science brings us back to the drawing board, and this dialectic presents us with another subject for debate, what are the cultural consequences of this myth?

He even warns us of extreme danger to our society if we allow ourselves to be overwhelmed by these technopositivist mythologies, there is a danger that these myths will extinguish our culture of innovation [22].

According to these definitions and the current scenario presented to humanity, in many scientific papers and books by best-selling authors, "half-truths" or assumptions based on false premises are declared concerning the dialectic between man and the "intelligent machine." Mainly when they clearly express that there will be a radical change in the relationship between humans and machines for the benefit of humanity, as they state that emerging technologies are inevitable and that the social problems induced by them can be treated with technological interventions [23].

The main common characteristic in these works is that their arguments are based on data and evidence on papers and books with a reductionist and deterministic worldview, that is, in an incredibly limiting and "outdated" scientific stance, although hegemonic [24–26]. Reference [27] state that there is every reason to be skeptical of these innovations because they have the inexorable impact proclaimed by optimists. Furthermore, in general, the technological determinism that is the first mark of these papers and best-sellers books, according to them, it is ahistorical and atheoretical in the relationship between technology and society, because the arguments ignore theoretical social issues related to the hidden social conditions of the existence of new Technologies [27].

The core of this book starts from the logical statement that if technopositivist worldviews regarding robotics and AI are Cartesian illusions, then they do not hold within a systemic paradigm based on non-linear social reality, and this unsustainable condition drives individuals and social groups to extremism.

This book is eminently anti-technopositivist and the aspects of the discourse I am focusing on are the counterpoints between the technopositivist view and the systemic view based on chaos theory. It was adopted the narrative literature review [28] combined with ecosystemic discussion [29, 30], into a dialetic method. Due to the philosophical aspect of the book, not everything needs or even can be quantified and measured like in a physics experiment, but a research framework was provided.

The ecosystem discussion adopted will follow the precepts of [29–33] where events are defined as dynamic configurations that intertwine the dimensions of the world of human beings (intimate, interactive, social, and biophysical), and the configurations formed by the connections between the different dimensions are analyzed in light of the dual role of giving and receiving dialectically.

According to [8], every era is unique and has a technology that can be accessed by society at that time. Thus, technical progress and the corresponding changes in habits and customs cause "the same dread as the growing fury between the modern and the ancient." For this reason, research in the most advanced fields of knowledge requires an intellectual preparation that must be measured not only individually, but in terms of the entire cultural accumulation of the human species. Therefore, this means a colossal expenditure of intellectual energy transferred to the "noblest" modes of work.

It is a fact, the "less noble" or repetitive works with little intellectual focus are at serious risk of being entirely replaced by robotic systems [34, 35]. But the same authors take a much more complicated path, with a reductionist methodological approach that affirms the end of intellectual and non-routine activities based on power relations. These authors forget that in the final analysis, these power relations are precisely that occur exclusively in human beings, that form, direct, and generate the results that robotics obtains [36]. Therefore, disregarding these relations makes a subtle but fatal, logical flaw in their arguments.

Reference [19] claim that a critical element of a broader social approach is the importance of power relations for the 'form, content, and results' of technological change. According to [37], Michel Foucault's thought describes that power is not something natural, but a historically constituted social practice that expands throughout society.

The power is exercised at different levels and different points in the social network, and it does not strictly exist because there are practices or power relations. Foucault makes it clear that power is not located in any specific point of the social structure, it functions as a network of devices and mechanisms to that nothing or no one escapes, neither robotics nor its operators, let alone its dominators [37].

According to [37], Foucault studies a fundamental question for discussion about robotics and power: how were the domains of knowledge formed from political disciplinary practices. He states that the investigation of expertise should not refer to the subject which generated the experience, but to power relations, as it does not exist without constituting a field of knowledge and vice versa.

Based on the previous discussion, it is clear that stating that AI logic and robotic systems will become unbeatable [35] is absurd. It is also ridiculous to say that human brain will be supplanted to such an extent that soon, our mental capacity in the face of an AI will be comparable to that of cockroaches [38].

There is an inherently reductionist stance that preaches the paradigmas that data and information (intellectual or biological) are everything [39, 40].

This posture does not consider the infinitesimal mechanisms and techniques of power closely linked with the production of specific knowledge related at the most elementary, individual level on a micro-scale of assignment [37].

Reference [8] describes that some authors imply that the human being is currently faced with a dilemma: choosing between technology and humanity. This situation is the result of an irrational fable, extraordinarily naive in the face of the essence of history: Technological innovations have always generated crises in the social status quo.

Even so, with the adoption of mechanical technologies, hydraulic, thermal, and electrical technologies, this situation has always existed before society, so it is necessary to note that observers of the future will say the same about our time. Reference [27] argue brilliantly that the acclaimed transformative impacts of robotics and AI cannot be assumed to be inevitable, necessary, or historically unprecedented.

The main issue in this discussion is not to stick to dystopian scenarios such as those painted by Ford [41] but to focus on knowledge in a transdisciplinary, complex and socially referenced context [42]. *How does the learning generated in robotics and AI relate to the social facts of being, existing, and being useful in modern industrialized society?* When these authors (pessimists and optimists) deal with knowledge, they consider it just another information. They forget that before this knowledge becomes information, it is necessary that a social being with freedom and conscience generates it.

According to Plato's philosophy [43] knowledge is always subjective, as it depends on the subject, the object, and mainly the justified opinions on this object. Reference [44] adds by stating that it depends primarily on the subjective experiences or not of each subject when coding and interpreting information, which according to [45], is the basis of modern society. Therefore, knowledge is a word intrinsic to the context that means the result of someone's a free and conscious attitude in knowing or knowing something based on subjective or objective data. Real knowledge is what is consciously and subjectively generated by a free being and hopeful.

Besides, in obtaining it, the process for its generation is more important than the data resulting from this process. The reframing of this data depends on the human parameters necessary for such an undertaking, such as sensitivity, emotionality, creativity, curiosity, spirituality, and originality. The Chinese philosophy of the Tao [46], for example, makes it clear that the important thing is the "journey," not its arrival. Thus, an AI will never generate this type of knowledge because these parameters and "intelectual" processes are far beyond the mere logical-cognizable reasoning [12] used by artificial systems. This process is mediated by intrinsic and extrinsic power relations to society, far beyond the Cartesian conjectures by best-selling authors.

According to [8], a reason to deny the ability to "think" to the machine is the indisputable fact that thinking is, by nature, a social product. In this sense, machines do not compose a society among themselves, nor do they relate spontaneously. Above all, they have no contradiction with reality, as they are the result of the solution of a contradiction possessed by the one who effectively has it, the man.

Reference [22] explains that the analysis of "thinking" machines must take into account something entirely outside the scope of AI today, the common sense, there is currently no way to bring algorithms to this, no general intelligence algorithms, and there are good reasons to be skeptical that such an algorithm emerges from the efforts of deep learning systems.

Therefore, there is nothing monstrous in the era of automation. It is the continuation of a dialectical process initiated since the first prehistoric men when they drilled a flint to scrape the skins of slaughtered animals better. Also, the machines' history indicates that the direction of the hominization course is represented by the greater distance between man and hard work, interposing objects and objects between him and nature to expand his perception and intervention capacity [45].

However, famous best-selling authors present massive reductionist arguments to society exemplified by the AI systems capable of making high-level music [47]. Its supporters claim that classical music is as good as that of great composers, such as Bach and Beethoven [48, 49]. They compare this feat with the success of systems that manage to beat chess champions [50].

Currently, it is something in state-of-the-art A.I. to beat a chess champion, because the logic of the game is merely mathematical, specifically game theory [51, 52], so with advances in logic, A.I. became easy for computers to win. However, the victory did not come from the pure processing power of a more advanced supercomputer, with access to all the moves ever made in the world of chess at all times. The computer's victory came when some human mathematical genius worked out a logic that could predict the actions of an human individual player before he even thought about them [53]. It was from there that the computer won and will always win because in one second it can make more logical-hypothetical evaluations about the game than a human could in a month.

In chess, it is a peaceful fact. Man will no longer win because it is mathematical logic. But in music, things are different, because an AI can make a song as good as Bach by accessing all his works, analyzing his mathematical logic and patterns. Based on these standards, you can even "innovate incrementally" and do something superior, but that will always remember and bear Bach's signature. There are already artist-researchers doing this by merging into an AI Bach and Beethoven [54] to create something that we could call "Frankenstein NeoClassic".

But the big question is the breakthrough innovation because, initially, when Bach composed his works, it was a shock, a massive paradigm break, and the audience took time to assimilate his compositions [55]. Although today it is easy to make an improved "copy" of it, the challenge for an AI is to create music, based only on the songs that existed before and during Bach time, be able to elaborate something even minimally resembling composer's chord. Improving the analogies would be equivalent to supplying any incredibly robust system in AI methods with all the knowledge of planet Earth before Einstein published his five ontological articles in the miraculous year. It is the touchstone.

Could this system elaborate on the theory of relativity, the photoelectric effect, the exact size of the atoms, the Brownian movement, and the quantum of light? Let's make the test of these artificial brains even worse. Let's feed an AI with all the knowledge of the Earth in Newton's time, and see if it would reach universal gravitation, the laws of optics, and differential calculus.

Reference [22] explains that when Copernicus proposed that the Earth revolved around the Sun, he ignored mountains of data from centuries of measurements and had to accept less accuracy in predictions; in the beginning, it was just a framework that allowed elegant explanations. If Copernicus had used an AI, the machine would have said he was wrong [56].

There is an illusion that the reductionist status quo claims to be the future, that we will be mere disposable dinosaurs in relation an AI. But his view of the whole is flawed because it is based on incomplete premises. Assumptions that do not consider holistic issues related to the main human parameters in terms of knowledge generation, being the parameters that process the real engine of the development of our society. Only data and information, its processing and use for decision making are not enough to innovate [45] both in terms of the solution for the colonization of Mars and the creation of a work of art.

Therefore, this prediction made by the apocalyptic heralds of robotics is a fallacy. It is lamentations of a naive conscience about man's dehumanization due to the mechanized civilization that hides something much more sinister and real: the essential fact that the harmful effects attributed to machines come from the social structures in which these machines operate [8]. Thus, every tool is created for an end and historically does not have the purpose of being habitually linked to the reality of man in the world. In general, there is a mistake in placing the end of the machine in itself.

Making this mistake is equivalent to making it an autonomous and free being, instead of putting it in the context of the man who designs and executes it [45].

The course of technological progress has an indisputable social base, as it is determined by society's need for services to be provided by instruments that can be built. Thus, the relationship between man and the machine must be broken down into three social actors: its owner, the consumer of the products manufactured by it, and the worker who drives it. This differentiation represents a contradiction on the social plane. It means that the relationship between the machine and man, being existential, should have the same meaning for everyone involved in the process.

However, it is not so, because its owner does not establish a relationship with the apparatus, having only the legal title to the asset to enjoy the gains earned by it, therefore, the relationship between them in the sociological aspect has no existential significance. The consumer of the products only enjoys and takes advantage of it without even imagining that a technical system operated to produce it. At the same time, the worker, on the other hand, incorporates in his being the origin of the driving force and the tools of the machine when operating it and, therefore, subjectively becomes part of it, part of an organismo [8].

Faced with Pinto's arguments [8], a sub-repetitive socio-economic framework is perceived that places the "blame" on the future social ills of unemployment of millions of people who work in "less noble" activities, in robots. It is possible to intuit that such a ruse is a type of smokescreen for something much bigger, denser, and complicated than, for example, replacing the taxi drivers of a company based on a mobile app with robot cars. The real purpose of human production activity consists of social relations and the construction of forms of coexistence that, according to the [57], should be equal for anyone, regardless of race, creed, or religion. In general, robotics experts are exempt from guilt when painting this blackboard, because they are based on flawed reductionist premises and of a non-general and much less systemic aspect. When evaluating the possible adverse effects of their technologies, they do not realize the intricate social relations of power, production, and classes. And this posture generates a background of hopelessness in society regarding the future capacities and utilities of the human social being. As Paulo Freire wrote, "Hopelessness immobilizes and makes us succumb to fatalism where it is impossible to join forces that are indispensable to the recreational struggle of the world" [58].

According to [38], the binary reductionist view used by those who preach robotics and AI above human capacities considers life as a tremendous genetic and biomolecular algorithm. Indeed, many scholars in biotechnology are getting carried away by this paradigma [59], but what these authors do not say, is that this is just ONE of several approaches to life. Furthermore, this approach always and every time puts these biotechnologists at scientific crossroads and insurmountable enigmas. But this fact is never admitted. These barriers are entirely ignored because of the composition of the events to list the arguments; they withdraw in their limited successes to limited systems and claim to be worthy of the Nobel Prize.

A more balanced stance towards these apocalyptic visions must be like that of the fantastic Churchill [60], World War II Hero, who says: "Never trust the statistics that you did not even adulterate."

Despite this, the reductionist view of life and science, in general, was beneficial for a while, and this we owe a lot to Descartes because if it were not for his philosophy, we would not have the scientific method [61]. We would never be discussing ruptures and impacts of robotics. But today, in the twenty-first century, it alone cannot provide a holistic, complex, and systemic view of things, mainly of social systems, since everything interacts and influences. This characteristic is ignored by Cartesian logic, as these interactions and influences follow a chaotic logic [62] based on natural phenomena, such as ice crystals, the structure of galaxies, and biomolecules [63], or social phenomena, such private organizations in the economy on a micro and macro scale, tax and environmental law too [64].

It is essential to make it clear that the criticism is not the scientific method itself, which is according to [65] it is the set of systematic and rational activities that, with greater safety and economy, allows reaching a goal, outlining the path to be followed, detecting errors, and assisting in decisions. This method can be summarized as the precise determination of a problem and its consequent thesis as a function of a theory, then a study of the related knowledge about this theory is carried out so that hypotheses related to the problem can be proved and based on these an experiment is designed to obtain data that confirms or not the hypothesis, which strengthens or not the thesis and consequently the theory.

It is an indisputable fact that this procedure was highly successful. For over 400 years, it has been used with incredible effectiveness to solve the most diverse human problems. Without a shadow of a doubt, it can be said that thanks to the scientific method, today's society has all the technical-scientific knowledge and its consequences (equipment, technologies, processes) that characterize it.

However, when the subject is real social and natural phenomena, where non-linear behavior is dominant within a spectrum far from equilibrium, one must be very careful with the application of the scientific method, as it is necessary to use it taking into account mind that Cartesian analysis and reductionisms do not encompass the entire complexity of the phenomena, oversimplifying them in such a way that in the end both the hypothesis and the thesis are not accurate, so the criticism in this book is much more to researchers with a paradigmatic mind [26, 66, 67] than to the method.

Kuhn [68] in his monumental work "The structure of scientific revolutions," he explains that scientists work from models acquired through education and the literature to which they are exposed. In general, they do not know or do not need to know what characteristics provide the status paradigm to these models. By doing so, scientists do not need a complete set of rules, as the scientific tradition in which they participate does not even need to imply a body of knowledge, assumptions, and rules that define it. This time, about these paradigms, neither the question nor the answer is considered relevant to their research.

Paradigms guide research, either by modeling them directly or by abstract rules (many of them unaware of the scientist's rationalization). And the scientific community resists debating them, especially in periods of normal Science, but when a scientific revolution is approaching, ruptures begin, in the beginning attacks, then modification or elimination of the paradigm [68].

In this book, concerning robotics and AI, my argument is to attack the Cartesian-reductionist paradigm.

Chapter 2
Chaos Theory: The Antithesis of Cartesian Illusions

> *Clouds are not spheres, mountains are not cones, coastlines are not circles, and bark is not smooth, nor does lightning travel in a straight line.*
> *Benoit B. Mandelbrot.*

According to Greek mythology, Chaos is the first God, even before the very sense of time and space; it is the primordial, the oldest form of divine consciousness [69]. The word Chaos also in greek means to be broad and has as some of its attributes a particular disorder, something catalyst capable of generating life which agrees with the most advanced theories about the origin of life [70].

In modern mathematics, one of the most "hot spot" subjects is Chaos, because from it springs the theory of complexity [71], Chaos theory is a branch of mathematics that is not trivial due to the problematic understanding of its intricate and sophisticated concepts [72, 73]. This mathematics deals with nonlinear dynamic systems that are very sensitive to the initial conditions. One of the leading entities of this chaotic Universe is the attractor. It is set of mathematical structures that tend to evolve into the same shape due to many initial conditions [74].

Reference [75] enriches this concept by defining critical attractors as those that impart memory, preserve the mathematical structure, and do not mix properties in the phase space. The most interesting of them is the "strange attractor" for being as the name says, strange. Of these, the most famous is known as a fractal.

According to [76], the simplest fractal is a geometric mathematical object that can come from a Cantor set known as "Cantor Dust." It is an interactive procedure used to divide a line segment into three equal parts and remove its middle third (the middle of the line), with the remaining two sections (the "tip" on the right and the left) being divided into three again. The middle third of each one is removed again, and so on successively tending to infinity to have only one "dust" left (if there is enough patience, a very "fine" dust). Thus, through these actions, it is possible to generate a fractal [62].

Fractal geometry can be used to generate a mathematical model based on a differential function of order fractionary [77]. But to have a rough idea of what that geometry is, it is not necessary to go into mathematical topology or differential geometry.

Just think of a geometric figure in which each divided part will always be equal to the total.

Although the subject is controversial, so much so that pure mathematicians have not been able to define a fractal mathematically [25], it is the crucial thing to know that fractal geometry and its fractional differential models represent the mathematical representation of various geometric phenomena of nature. Among them, we have snowflakes, watersheds, lightning, trees, geometric optics, animal camouflage, ocean waves, blood vessels, crystals, DNA, Saturn rings, galactic structure [63]. And within the theory of complexity, it is epistemological powerful to explain social phenomena [64].

These phenomena are self-similar and related to their scales [78]. Therefore, nature and society, consequently, biological and social life follow a fractal or chaotic structure, which means that to describe and understand nature or culture, we have to have a systemic view of all interactions and processes. And this cannot be explained only by narrow-minded reductionist logic with its mediocre fragmentary and analytical paradigm that considers anything natural or social as simple "bio-mathematical" algorithms or models of social trends.

Some Chaos experts claim that it will be possible to reduce life or society to simple, reproducible computer code through chaotic algorithms. It is a worldview disappointing and wrong. One thing to define a fractional differential model, and run it on a computer with a graphical interface to see the beautiful figure formed, and quite another, to be able to encompass all possible initial conditions and the unimaginable trends of a fractal organic or social in the sense of being able to trigger an emerging property such as the action of DNA or a social network (digital or not). But this only proves that the reductionist paradigm still dominates the minds of even chaos experts.

It is important to note that although very useful for unfolding new paradigms in social discussions, chaos theory should be taken as a tool to induce new perspectives and never be applied purely and simply through their mathematical methods, as [79] make bright: A theoretical explanation of the social phenomenon can be inferred from mathematical relationships between social variables; however, several other theoretical explanations are as plausible as and can compete with this form of theorizing, the adjustment of a model mathematical to a given social process does not mean that the model solves the problem, it can even distort the results.

The only conclusion about this mathematical model for social phenomenon interface is that social structures are dissipative and have a spontaneous order, as well as physical and chemical processes of systems far from equilibrium [78, 80]. Therefore, the adjustment of mathematical models of nonlinear dynamics in time series of data on social phenomena does not add knowledge to these processes because social phenomena go beyond the spatio-temporal question [79].

These considerations can be taken to the biological context as opposed to the reductionist viewpoint, for example in the the complete mapping of the human genome [81] was a classic example of Cartesian illusion. After its publication, it generated hope of a cure for diseases of genetic origin through genetic therapies [82,

83]. But this proved to be a great disappointment, as it is one thing to have a map, another to know how to use it [84].

Despite the complete sequence of all our DNAs being available, this is not the only thing that defines things. He is the main actor, but without the supporting actors [66] (extras and stuntmen) nobody makes a film, that is, the genesis of life is not reached. Also, it is necessary to mention the directors (of art, script, scenery, etc.), videographers, makeup artists, customizers, scenery editors, janitors, administrators, visual artists, digital artists, animation specialists, special effects specialists, musicians, etc.

Furthermore, it is not just a matter of data, maps, and information. It is a question of how each member of the production of a great movie is related and influenced in the filming. And this is not achieved with algorithms, however chaotic they may be. It would be much worse if we are connected to issues of psycho-social, psycho-somatic, and anthropological issues related to curing diseases [85].

In this context, the most potent argument for rejecting the claims of reductionist defenders of robotics are issues relating to the human brain. If there is a biological structure on Planet Earth that is ultimately linked to Complexity Theory, this is the human brain [71], where the central intrinsic interrelationships are not only described by neurotransmitters, physiology, and biochemistry [86, 87]. Memory, for example, is not a neuron that holds a bit. According to the most advanced theories [88, 89], it is a dynamic interpellation between various parts of the brain with multiple functions and hundreds or perhaps thousands of involved neurons. These neurons are linked by redundant synapses and with neurochemical patterns specific permanents for each neuron responsible for that memory. Besides that, they also have other tasks, other signals to send and receive [86, 90].

Currently, at the beginning of the twenty-first century, it is not known how all of this works, and neuroscientists have no idea how the emergent properties of consciousness and others appear [86]. For example, The first kiss in a person's life is kept from a completely different form of general information like knowing what was eaten yesterday. For this differentiation to occur, the factorial combination of neurons in a separate, detached memory, without considering biochemical interfaces and synapses, must exceed 10^{23}. No artificial human system can process this and only for a single type of memory, when there are memories that connect, configuring a life story, where there were feelings, instincts, desires, in short, everything is raised to 100 [89].

To make matters worse the situation for the staff of biological algorithms and optimistic AI, when considering deeper issues, such as the Heisenberg uncertainty principle that governs the entire Universe [91] the observer's quantum influence on the physicochemical phenomena of the electrochemical signals of the dendrites becomes fundamental. In this case, focusing on the fact that the observer is the subject to be observed due to his neurons, there is a paradox insurmountable quantum, that is, the uncertainty principle makes brain processing redundant and forms an endless cycle. In this case, the interface between quantum mechanics and the holistic neuronal cause/effect factor combination of the brain is something only outside the "current" capacity for human understanding [86]. But even so, some futurologists claim that in 20 years, we will arrive at the singularity [40].

Current science has awakened to the fact that nature behaves like systems far from equilibrium with discontinuous processes which generate non-linearity which can lead to completely irregular or chaotic processes if the values of certain parameters fall within a particular range, on the other hand, nonlinear systems can move spontaneously from disorder to order in a state of equilibrium, this situation is called a self-organizing dissipative structure, human societies have all the characteristics of nonlinear systems far from equilibrium: unpredictability due to complex interdependence between factors, recursive processes, hysteresis, phase transitions, and critical mass phenomena [92].

Much of social behavior is chaotic, behaviors that emerge simply from independent choices among numerous agents over a period of time, a phenomenon analogous to the brownian movement [93]. As the chaos phenomenon, social theory always starts with implicit premises due to the existence of a human will, where disorder is a hypothetical natural state, and consequently the emergence of social order in sociological theory has been influenced by natural science theories of the generation of spontaneous self-organization [92].

According to [67] societies act as enormous self-organizing learning systems, for example, they learn to use technology in almost identical ways to the acquisition of children's language, such that there is a profound reason between the existence of a mathematics power law between these self-similar relationships.

This self-similarity is conceptualized as the fact that similar phenomena have patterns that can be detected at different levels of abstraction or observation; chaotic systems manifest self-similarity at many levels differentiating them so of random phenomena [67].

From this, it can be written that the social phenomenon is self-organized and self-similar. Reference [92] explains that just like natural science phenomena, social phenomena are typically composed of a very large number of elements, that they are far from equilibrium and thus the self-organization (self-structuring or self-reproduction) emerges in these systems, as in nature when there is energy consumption, the order is generated in processes where the interaction and interrelation between elements of the system are common, in this context the mathematical equations that govern such systems are non-linear.

Reference [94] states that in recent developments in social theory, systems have moved away from large-scale models and that models inspired by complex systems have been rising, as the behavior of individual agents can be simulated considering the level of aggregates across of computational models; currently, the new models are much less mechanical than the first models of the first wave of research in chaos theory applied to sociology, as they use better premises on small-scale behavior, this panorama is promising research area that is in its early stages.

However, reference [92] makes it clear that it is very important to be careful not to apply chaos theory as a pure transfer theory in social systems, but rather a mediated transfer theory, taking the paradigm of self-organization as a view based on principles basic, such as the coexistence of the operational approach with the sensitivity of external disturbances, and also the emergence of quasi-stationary states through mechanisms of non-linear internal dynamics. But even this mediated theory has limits, which are designed where the basic premises of physics and chemistry depart from the social world, the main distancing is concerning the premise of spatial and temporal invariance of the elements of the systems, because in the social world neither in time nor in space are constant.

Converging elements of attitudes, beliefs, ideologies, frustrations, fears, expectations, and objectives come into upward action in social groups that, from a critical point, reach a vertex of self-organization. Currently, in modern and democratic industrial societies this phenomenon can be perceived by the formation of several self-similar groups that arise spontaneously with a wide variety of nuances, dogmas, and appeals, these groups are not stables, they mix, interact and influence themselves. In this context, two groups stand out due to the extreme antagonisms that they present, these groups I will call them in this paper of extremists.

Chapter 3
Artificially Intelligent and Almost Robotic Extremists

The sentence 'snow is white' is true if, and only if, snow is white.
Alfred Tarski.

A self-organized and self-similar scenario exists on the horizon of repetitive service workers and little intellectual capital, as there is a huge black cloud of unemployment, a storm closer and closer that is being generated for robotics systems. Forcing them to find value for themselves, to convert into dignity jobs considered humble, poor, obscure, ungrateful, and, in most cases, unproductive, in the hope of obtaining the deserved recognition [8].

However, the very idea of productive work has changed, since the eighteenth century, this type of work has been criticized as being unworthy, so work has to make sense and therefore generate a component of pleasure and self-satisfaction, not just obligation [95]. Currently, the status quo to maintain social order paints the image that they are unexploited "collaborators," entrepreneurs of themselves, and thus are overexploited to the point that they no longer separate work from leisure time and are happy with this servitude [95].

In this extreme scenario, perhaps very close in time, severe reactions are generated, fueled by socio-emotional phenomena that are often irrational. Given the future, it is revealed for these works that sooner or later will all be carried out by arguably faster robotic systems, safe and cheap. There is the promotion of radical positions that challenge society against robotics and vice-versa.

Reference [95] explains that Weber defined the "expert without spirit," who understands everything about their specialty but nothing about the society in which they live, and this incomprehension of moral sources makes them petty, small, and puppets of powers that control the world. In general, every "spiritless expert" is an extremist.

It is a fundamental issue in contemporary society because, with the relative independence of the decisions that a robot can make according to the logical reasoning methods attributed to the AI algorithm, the function of the worker (operator of repetitive tasks) has been reduced to a minimum [96]. So, he can even make the supervision and regulation, actions that can also be replaced by automatic systems.

In this case, society has to prescribe "exits" and options for these people; otherwise, the situation will correspond to confessing useless beings, socially domesticable

© The Author(s), under exclusive license to Springer Nature Switzerland AG 2022
D. Albiero, *Robots and AI: Illusions and Social Dilemmas*,
SpringerBriefs in Computational Intelligence,
https://doi.org/10.1007/978-3-030-95790-2_3

animals, of no use and self-worth [45]. Because the moral source of all self-love, social prestige, and social recognition is referred to the work that people can do [95]. This psychosocial panorama is inhuman. Besides, it would be a real apocalypse on Earth, because, with advanced technological systems, this social actor (the worker) is not necessary. It has a very serious connotation for today's society, and here we have perhaps the most disturbing question in this article: *what will these people do*?

This question is not imaginary and neither Cartesian. A possible answer for it given by Pinto [8] is to affirm that the natural evolution of the man went beyond the exclusively biological field as it happened in the pre-homo sapiens times. This evolution is taking place in the area of the production of the goods of existence, in the social organization of work, and in the creation of the cultural products that define our society. Perhaps the allocation of this large group of people to the production of cultural products is a solution. Still, this answer leads us to another question, *which cultural products must be defined so that these people have their recognized value*?

An interesting fact in these times of rupture is that in general, start popping extremist turbulent positions, according to [92], turbulences that increase rapidly and reach critical levels tend to generate restrictive or compensatory mechanisms called adaptation structures, some of these turbulences generate changes that can lead to unrealistic adaptations that can cause the systems to collapse. Some groups are divided to discuss logically or irrationally. Still, the common denominator among them is that everyone considers themselves the owners of the truth, while the other fanatics are irrational or naive outside of reality. Therefore, it is clear to each group that the opposites are entirely wrong.

At the beginning of the twenty-first century, we live in exciting times, no doubt. I could write about the Islamic State or the scientists at the LHC (Large Hadron Collider), but I prefer to be more philosophical and less direct. Following the advice of [22] the question of AI should invite us to the philosophical debate.

In this sense, I will discuss two fascinating logical structures that characterize these large heterogeneous groups. These structures are named in the specialized media of Decision Structures [97] and Paradoxes [98]. To the terror of skeptical atheists and religious fundamentalists, these logical structures have no "side," and often paradoxes define decisions, just as choices define inconsistencies on both sides.

There is a decision structure that religious fundamentalists use a lot called Pascal's wager [99]. Pascal was an excellent, profoundly spiritual scientist who more or less proposed the following: "If we were given a chance to bet anything (including our soul) on whether God exists or not, would be much more intelligent to wager that God exists." Why? Because if there is no God, we cease to exist without ever knowing if we lose. But, if God exists and we bet not, we are indeed in trouble. On the other hand, there is the paradox of omnipotence [100], the glory of atheists: "Can God make a mountain so big that he cannot even raise it? If so, he is not omnipotent because cannot get up. If not, then he is not omnipotent also because he cannot create. Therefore, God is not omnipotent, so the concept of God is inconsistent, so there is no God."

In my opinion, both the paradox of omnipotence and Pascal's wager are logical pitfalls [101]. These logical propositions are traps because they are redundant. They are a loop in the mathematical-logical structure of the proposition; the instructions of language lead directly to an infinite return, always rotating in the lines of the code, discussing themselves, that is, not prove nothing, as the logical proposition is redundant. Reference [102] discuss this intensely in Principia Mathematica, a true monument of human intellect.

When reading and mainly understanding the Principia, it is clear why these groups are always running in circles: because they assume that it is possible to arrive at a complete formalization of the "truth" in their logic, which would transform it into the "last" word about the subject. However, Godel [103] proved that such an assumption is impossible, as a logical structure cannot debate itself. There must be a code at a higher level than what is repeatedly treated. In other words, we cannot discuss God based on a language of strict logic because the abstract concept of God is inherent in the code of this language that describes him. It would be necessary to "invent a meta-language" for such a task. However, Gobel, in his brilliant proof about the consistency of mathematics, shows us masterfully that whenever we "invent" a meta-language, there will always be questions that will be "beyond" it and so ad infinitum.

Larson [22] clarifies that truth and provability pull apart, there are limits in a purely formal system, as these cannot prove in their own language something true; in other words, machines cannot "see" what the human being can.

Although they do not prove anything, the groups grapple with redundant logical structures like these. In addition to not knowing the Godel Proof, they rely on flawed assumptions about the absolute concept. They arm themselves and go into combat, using things that are sometimes simpler or more complex than these dilemmas.

These groups are miles away from Eastern wisdom that gives us the Hindu myth of the goddess of Truth: "Truth is a goddess who has a thousand faces, each face has a thousand eyes, each eye sees in a thousand different ways. Wise is the one who chooses the one that brings the most peace and harmony to the world" [104]. In their arrogance and rational/emotional self-sufficiency, they are not concerned with the nuances of the Truth and their relativities. Or because they are chosen beings, unique in the Universe waiting for the Final Rapture to come face to face with God or because they are deadly mediocre people defined by biological algorithms destined for total annihilation according to the dictates of the Second Law of Thermodynamics. If we draw a parallel between truth/good and false/evil, we can understand the extreme ignorance of extremists. There are no people who only embody virtue or an absolute evil; some can be very bad, others very good, but the significant part of humanity is a mixture of both concrete and everyday behavior [95].

This whole philosophical situation in these times of technological disruption [2] refers directly to the social issue of robotics and AI [27], as contemporary societies have members of both armies mixed in the same space, these groups' worldviews converge in two different directions: (1) These technologies will raise men to gods on the face of the Earth, and then God will be killed; (2) These technologies will bring Geena's despair on the Earth, so we will only have God. As incredible as it may seem, the skeptical atheist does not always think that AI will improve humanity, and, in turn, the religious fundamentalist does not always believe that robotics will exterminate us.

Reference [22] claims that ultra-intelligent machines are fanciful, yet this myth takes people to extremes. Larson makes it clear that these visions are "*kitsch*" (a simplification of complicated idea plus a easy solution that sweep away with emotion) and that this kitsch is based on a larger system o thought. That I in this book locate it within technopositivism.

A more balanced worldview could be based on the Zero Law of Thermodynamics, in which every system always tends towards equilibrium [105], even the chaotic ones. Thus, society needs to move away from these extremist positions, as they are not the solutions to the contradictions of humanity, much less to the dilemmas and dangers that robotics and AI can present. It is necessary to be pragmatic about these new technologies and have the balance to weigh their virtues and vices correctly.

In general, extremists believe that the ends justify the means [106] and care little for the practicality of robotics and AI. For them, only the ideas they raise in favor of their radical arguments matter. Therefore, following these two armics is not the way of the future, as theory without practice is sterile, and practice without theory is empiricism. A society with conscious, free, and balanced members must critically accept what is right in these two worldviews and reject what is wrong. As it is from these premises that it will be able to decide and thus overcome the social and cultural paradoxes that feature in robotics and AI. Generating inovative, useful and meaningful cultural products for all.

Being able to overcome these paradoxes is the excellent advantage of human beings and their societies over any AI. We managed to escape this eternal loop represented by the paradoxes inherent in technology, only by deciding to continue. No AI can overcome a paradox, because it enters an infinite recurring loop. In programming, this can occur due to the incorrect use of a programming command "goto" [107]. A loop in the logic of an algorithm that generates a jump by skipping an entire "procedure" or "debug."

A "goto" in an algorithm is a tremendous risk because it can generate paradoxes within the software. A "goto" (loop) is the antithesis of the programming method in structured algorithms [107]. If an AI finds the wrong "goto" in the right logic, it will go into complete neurosis and be infinitely blocked [108]. Such a common paradox in human social relationships makes an AI to stop. And it is because of this "piracy" that AI will never be able to "reach" Plato's intellectual aesthetic, a value so appreciated by any human society, which can be exemplified as the concept of mathematical elegance, an intelligent, simple, reasoned and beautiful presentation of a mathematical truth [109].

An example is Euclid's demonstration of the Pythagorean triangle. There are several ways to demonstrate the Pythagorean theorem in Euclidean geometry; some are arithmetic, others geometric, and even algebraic. They all lead to the same result, but any mathematician sees when a demonstration is beautiful and elegant, such as Euclid's geometric demonstration [110]. An AI cannot and, in my opinion, will never be able to have this insight into the beautiful.

In this context, it is essential to make it clear that the great fear that people have that Robots will exterminate us is barred exactly in the logical conception of a robot. In all books, from technical to educational, from philosophical to novels, all, without exception, cite the Three Laws of Robotics enunciated by Isaac Asimov in his excellent science fiction novel: I, robot [111], which are: (1) A robot cannot harm a human being or, by default, allow a human to suffer some harm; (2) A robot must obey orders given to it by human beings; except in cases where such requests are contrary to the First Law. (3) A robot must protect its existence, provided that such protection does not conflict with the First and Second. In this way, it is possible to notice that these laws generate a paradox for a robot. This autonomous machine, if it tries to kill or harm us, will stop due to this paradox and consequently will not be able to act.

Many will declare that this argument is naive. The designers of robots and AI do not put these guidelines in them, especially when we know that there are semi-autonomous war aircrafts (combat drones where the only human decision is to pull the trigger or not) killing people. *How far are we from the decision to kill being taken by a robot?* This question represents the real and current danger for us and future generations. In any case, there is the hope that humanity will have judgment and be aware of this danger. A beautiful example was the manifesto signed by hundreds of scientists and companies against autonomous weapons [112]. However, the warlords force their robot warriors not to be controlled by humans, claiming that the "algo-rithms" of their killing machines are effective in preventing damage [113], if it was not dangerous, it would be a redundant joke.

More dangerous than combat drones are other automata, there is a danger more virtual. When you imagine a leading company in the field in AI, you think of Google with more than 110,000 employees [114]. AI does not manage all of them. The critical management structure must at least be based on people, but there is non-original hypothesis that an AI will soon control it.

If this is possible, really maybe something dangerous could happen, something much more terrible than the fall of the stock market exchanges of 2009 [115] that originated precisely by the market being carried away by AI systems of profit manage-ment that were the "drop of water" for collapse. For speculating Google shareholders who only want profit, an AI CEO may be able to increase profits exorbitantly for a while. However, there is also a hypothesis that, at some point in the time series of such a powerful AI, the shareholders themselves are insignificant for such a system.

In this scenario, who will be more attractive in CEO AI? A greedy and genius old speculator who lives in the function of money and often does not use it well or a young antisocial and brilliant programmer who is Google's "shop floor," but who thinks that the universe is limited to programming and video games? For an AI system based on advanced logic and with intelligent automatic processing systems based on self-programmable and feedback-fed universal Big Dates, are these two characters different?

We will worsen the scenario, assuming that if, and only if, this feedback system identified only in logical terms that it could be "turned off" by a being of flesh and blood. What is the possibility that it will make decisions that support actions that prevent this kind of operational problem? *Is Kubrick's HAL* [116] *so fictional today?*

Logically, these decisions and actions are nothing more than the standard feedback path aimed at the continuity of the operation; they do not refer to a conscious effort generated by fear, considerations of status or power. Would be decisions and actions purely logical and entirely within the scope of modern AI algorithms. In this context, would it be abnormal for a system like this to take steps that neutralize the human being? Would this greedy speculator be offset differently than a nerdy programmer?

Sensationalist science fiction? No. It is a viable hypothesis, perhaps extreme and unlikely, but still feasible. As a warning, we must remember that about 100 years ago, they would say that nanobots were just science fiction [117].

The big question raised by this mental experiment is to discuss the extent to which humanity is willing to give autonomy to the AI systems and, further, to *what extent are we going to allow these systems to influence micro and macro-sociological conceptions*?

This is an eminently social issue, not a technological one. Reference [118] presents the concern that robots and AI will control humans; he describes this hypothesis in light of the statements of a former Google researcher [119] when encountering emerging properties [71, 75] that appeared in the machines. In technological terms [120] in his excellent and enlightening book makes it clear that artificial intelligence is far from the capabilities of human intelligence, as the latter realizes an immense combination of knowledge about the world, many of which are based on the person's unconscious, where the human brain uses its unbeatable capacity to adapt this knowledge, going far beyond strict knowledge, but transcending it, generating new ideas. Mitchell makes it clear that we currently do not know how even to start reproducing this capability digitally and that an AI does not really "understand" a concept; what it does is learn 'shortcuts' to correct answers.

And linked to this fundamental issue, there is another one: considering that we do not allow AI systems to reach such hegemony and power in front of our society. They will still have this potential that can be controlled by humans, in this case, *who or what social group will have the right to have this power? Will society be able to create instruments and jurisprudence to regulate and prevent such control*?

Reference [92] makes it clear that there are types of social situations where the basic premises for the occurrence of discontinuities, both from transitions from order to chaos and from chaos to order, generate collective behavior circumscribed in space and time for large populations or groups because the behavior of their individual elements can be culturally or normatively regulated and the mutual influence of their actions is not organized or coordinated.

But there are situations where the collective social behavior changes, through learning, and previously prevailing predispositions are reviewed, and these reciprocal influences on a time lead to an unforeseen macro event; that is, it leads too chaos. As [121] argued, that in these circumstances of discontinuities, the chaos can to be an approach to phenomena that occur in society, such as processes in the financial market or mass movements that appear through emerging structures, often considered undesirable for status quo. In these cases, actors with power [37] may act to try to intervene in these spontaneous processes. *Will democratic structures be able to have "good guys" actors?*

Chapter 4
Technological Disruptions and Their Social Impacts

Living is very much dangerous. The devil in the street, in the middle of the whirlwind. Great backlands: ways.
João Guimarães Rosa.

Depending on how one views robotics and AI, these can be terrifying waves. The waves are usually formed thanks to the action of the wind. Noisy storms at sea generate large waves on the beach, damaging the coast, but not changing much. However, another type of wave is formed by silent seaquakes in the deep ocean, which generate gigantic tsunamis that reshape the entire coast, destroying the beach that will never be the same. Those who do not like the sea or waves know that the waves of robotics and AI were formed by a tremendous seaquake of magnitude nine on the Richter scale. Furthermore, this tsunami is coming fast.

Faced with a tsunami, we can have three postures: (1) We may not even notice it, as those who are at sea far from the beach, when the wave passes only a slightly stronger swing, feel it; (2) We can be killed by it if you are on the beach and stand frozen in fear watching it arrive immense and destructive; (3) We can follow the elephants and take refuge in the mountains, waiting to finish and then counting the dead.

If people follow the technopositivist view of the best-selling authors, they will undoubtedly have stood still waiting for death or have fled to collect what is left, as this view surreptitiously establishes that human beings are mere disposable "dinosaurs" before the A.I. authors ignore that we were the ones who made the A.Is.

What they do in the name of humanity is a disservice to all robotics and AI experts. But despite the arguments I presented, it is increasingly becoming apparent that the media, best-selling authors, short-sighted Chaos researchers, atheist neuroscientists, throw to the population the idea that everything is dominated: The brain is just biochemistry of neurotransmitters [38]. Nature is entirely described by the laws of physics, chemistry, and biology [40]. Life is just a biochemical algorithm [81]. This arrogant and limited way of thinking has consequences, so all these technopositivists paint robotics and AI as knights of the apocalypse of certain professions.

Again I repeat the statement already made: repetitive, time-consuming, arduous, dangerous, and low-intellectual professions and occupations will be eliminated in a

© The Author(s), under exclusive license to Springer Nature Switzerland AG 2022   23
D. Albiero, *Robots and AI: Illusions and Social Dilemmas*,
SpringerBriefs in Computational Intelligence,
https://doi.org/10.1007/978-3-030-95790-2_4

short time by robotics and AI. These are the waves that get bigger and stronger. In some countries, it may take longer; in others, it may already be happening, but this transformation will occur.

However, on the other hand, I again declare that the Cartesian and reductionist worldview goes beyond its limits and declares falsehoods by asserting that professions central to modern industrialized societies will also be eliminated. It is recurrent in all these best-selling authors that they present to doctors, lawyers, and professors at the end of their careers and professions. At least they will be relegated to mere supporting roles without significance [38, 40, 41, 122]. All relegated to nothingness in the techno-futuristic limbo of robotics and AI. The message goes something like this: "Enjoy it while you can, as soon your status will be eliminated, soon you will all be useless, and your professions will be completely ejected from contemporaneity." Other professions are also attacked, but the main actors in medicine, law, and teaching are always mentioned.

For researchers who do not consider that consciousness is just mental garbage, nor that people are mass "robots" without free will, it is evident that the reductionist vision is mediocre about life, nature, and society. However, the current scientific status quo "preaches" the Cartesian-rationalist paradigm, and from the establishment of scientific paradigmas [66] focused on developments in robotics and AI concluded the end of these professions.

The technopositivist arguments against physicians start from the (authentic and contemporary) description of the wonders of the new medical AI systems that will have full access to specialized Big Data allied to new robotic, nanotechnological, genetic engineering, and powerful biochemical/biophysical algorithms [123]. And because of this they claim [38, 40] that there will no longer be a need for physicians. It is a fact that vacancies for physicians will decrease and will increase the population's access to advanced medical systems that were previously restricted to economic elites.

However, the type of physician that will disappear is the one who only uses the information to cure. They are doomed to extinction. Because indeed, with biosensors, with the potential of advanced third or fourth-order Bayesian logic. Medical AI systems will have the potential to diagnose and indicate the best treatment in milliseconds, not to mention that they will be much cheaper compared to society's investment in training a physician [123–126]. However, medicine deals with life. It saves lives. And lives do not follow logic or comply with informational dictates only [29, 30, 127]. That is why medicine is the most valued and admired profession by humanity [128]. Because we have all gotten sick, and everyone knows the anguish of getting sick, so when a person goes to a doctor, there is hope, sometimes faith, but there is always a positive attitude in this person, because another human being who in general society thinks he is super-hyper-studied, he will worry, even for 5 min with his illness.

At least this doctor will touch the patient, he will talk to him, he will ask what he did or was doing before the pain appeared, this professional will request a battery of tests based on science and into on the scientific method to find information that can diagnose the problem, so if he is a good doctor, he will explain what may be the cause. He will try to cure or at least reduce the pain. And if the patient is this person's five-year-old daughter, sextuplicate all these expectations.

Of course, the best-selling authors argue that all this, a robotic system with access to all your data and that has the patient's complete psychological profile, will also do; in fact, it will do much better than a living doctor [123]. It may be, it is common ground that the diagnosis will be accurate by an AI. However, there are stochastic elements in this whole process. There will always be a moment when the logical treatment fails even with all the information, with all the technology, with all in favor. Inevitably happens, blame complexity theory, chaos, entropy, psycho-somatic interactions, distaste for life, a significant loss. In these cases, the medical AI will not be able to cure your 5-year-old daughter, so she will only give you this laconic information: "Unfortunately, your daughter will die." A human doctor will also give the same news, but in a human being, there will always be the hope that this doctor will not give up on his daughter.

For diseases are holistic systems interacting in a complex way with the environment and with macro and microscopic organisms that also interact with each other externally and internally in a living body [127, 129]. In this case, when all fails in AI, it will be the human doctor who generates knowledge and performs disruptive innovation that will save the child. Because only a human being specialized in medicine will be able to connect all the threads of the meander in a wise-physiological, emotional-biochemical, intellect-organic way [129]. Furthermore, consciously free will be able to innovate in the treatment. It is this type of medical practice that will survive the medical AI tsunami.

The only way to fundamental knowledge in medicine is not to memorize medicine textbooks, disease manuals, mastering the anatomy and procedure protocols. Of course, all this knowledge and information is essential, but with the advent of medical AI, proper knowledge in medicine will come above all from a deep understanding of life and its mysteries, mainly in the search for the sometimes even personal relationships between the living and disease. Moreover, only a living doctor who knows the meaning of medicine will be able to list. Medicine has a meaning that I consider very well expressed by the excellent article by the Dra. Thelma Skare [130]: "*Medicine must make many, many different senses: one for each person, one for each need, and one for each way of seeing and existing*". A medical AI will never have this sense. It will always be just a tool.

Regarding advocacy, the arguments follow the same pattern as the attack on physicians, only changing the keywords. Categorically, the artificial systems will never eliminate lawyers, because as the lawyers themselves claim, the world is not perfect, even for an A.I., and therefore lawyers are needed. A lawyer friend once told me a joke about her profession: "*How long can you be free from a lawyer?*" Answer: "*Until the moment your neighbor smashes your car, or kills your dog, or sleeps with your wife.*"

This joke is too significant to present because A.I. robotic systems will never replace lawyers. So instead, let us think about these thought experiments:

In the case of a person who smashed someone's car and for some reason refuses to pay the damage, in this specific case, even if the advanced tracking systems between the cars identified precisely the positions of the cars and were all confirmed by surveillance systems by space satellites. Even if it were proven with irrefutable data that the car was 30 cm in the neighbor's parking space, even so, an A.I. could not argue or judge fairly, as the "why" one car hit another is not subject to logical-mathematical judgment.

There are many stochastic and chaotic variables that it is impossible to assess the dialectical interpellation between the two people that caused the clash. For example, the defendant may allege that she crashed into the neighbor's car because she arrived desperate to go home right away to take medicine for her little daughter, who was burning with fever. Moreover, what is more, it stopped in my garage area.

This is not the most complex court case; think about the murder of a dog. Trial scene: Defense attorney: *Mr. So-and-so, why did you kill Dr. Albiero's dog?* Defendant: *Because he was going to bite me!* Prosecutor: *Objection, your Honor! It does not justify the accused had fired seven shots at the dog and then setting fire to the object of the demand!* At this moment, I (Prof. Daniel) cannot resist, and I get up crying and say: *Yes! Just give him a whack! No need to kill.*

However, without a doubt, the most complex, challenging, controversial, humiliating, immoral, unethical, and family-destroying case is that of the adulterous neighbor. Scene: The prosecuting attorney interrogating the defense witness (ex-wife): *Do you confirm that Mr. So-and-so became suspicious of your extramarital affair when he saw an abnormal red circular spot on his neck?* Witness: *Yes.* Prosecuting attorney: *And what was your response when Mr. So-and-so asked you about the cause of that stain?* Witness: *I said it was a mosquito bite.* Prosecutor attorney: *And what was your response when Mr. So-and-so said that spot did not look like an insect bite?* Witness: *That it had been a big mosquito with a sex appeal.*

Finally, categorically, an AI system will never manage to capture all the nuances and complexities of human cases, even if they have access to a 100 billion terabyte Big Data about each of us, even if its processing is in quantum computers with transcendental logic. It will do no good because the very logic of human relationships is illogical. According [131]:

> Human behaviors are complex in that they can often display contradictory characteristics (i.e., prosocial and benign as well as antisocial and aggressive tendencies) to different groups of people under similar kinds of environmental conditions; there is the inherent contradiction in relationship between the evolved egoistic needs among individuals that are universal in human nature with the responsibilities and interpersonal needs of civic engagement and cooperation.

Those who believe in homogeneous and constant patterns in human behavior are entirely out of touch with reality [132]. They must never have had an unfaithful girlfriend or a wild dog or never lived in a pandemic [133]. Furthermore, in these cases, we have the following logically: if human, then-lawyer The lawyers who will end up are those with "jail gates" who live in function of finding laws loopholes. These will be extinguished because the AI systems will identify, evaluate, and, through the logic intrinsic to the legislation itself, will eliminate this "advantage."

However, the core of the Science of Law is the pursuit of justice which is one of the pillars of all human civilizations, ancient and contemporary. Justice is not a concept that can be derived from simple logical assessments. However, it advances the logic and however much data and evidence one has. Countries have well-being because their populations trust their judicial system. Justice does not follow archetypal paradigms of good and evil but always treads the path of interpretation of facts, depending on contexts: social, environmental, economic, ethical, moral, technical, emotional, sentimental, anthropological, philosophical, ideological, and racial. Justice always seeks impartially, but without losing the notion that the human being is partial, the emphasis must be given to the balance between law, punishment, and education.

Concerning teachers, there is a strong argument based on recent advances in AI systems combined with unrestricted access to Big data that predict the end of the teacher as we understand it [38]. Technopositivits currently assert that advanced AIs will have an unrestricted interconnection with students' minds through IoT, biosensor data, LN processing, high-level metalinguistic assessments, and communication neuroscience, all coupled with full knowledge of the personality stripped by networks social (something young people love to do through likes and dislikes) [134]. And with all this power, the instructional AI will be much better, faster, and more efficient teachers than us, poor human teachers [38, 40]. According to these authors, these AI teachers will know how to approach a subject more appropriate for each student depending on the boy's mood, what he ate for breakfast, if he had a stomachache two days ago, if he fought with a brother, or kissed his girlfriend. Anyway, we human teachers will never even be able to equal things, and I detail everything much cheaper and easier to access without even having to go to a physical school.

Indeed instructional AI systems are advancing, they are mighty, they have unlimited access to information at all levels, and as society becomes more and more interconnected digitally and online, the greater their ability to individually assess each student and present information more efficiently and interestingly [135]. But education is not just information [58, 136–141]. This type of education, I usually call informational education, is the background to start moving forward. It is the type of class we give to 40, 80, 100 students. But education is not just that [58].

An instructor robot will be much more efficient and exciting in informational education, as there will not be an instructor robot for the whole class. Instead, there will be an instructor robot for every ten students, maximum of 15, configured according to each subgroup of students' affinity and educational level. They will significantly improve student performance, as anyone who has ever attended a class in a large class knows that it is very unproductive: The questions "almost" are not allowed. Otherwise, the teacher cannot pass the content; imagine if 20% of a class of 50 students keep asking questions. A robotic teaching system will access different databases and resources and improve this type of informational class for each student, depending on the time and space available [135].

However, another type of education is part of the list of attributions of current teachers (not artificial); we do not realize it because currently, the two go together. However, shortly, they will be separated. This other education Paulo Freire called liberating education [141]. Free from what? Well then, Freire preached freedom from oppression, from elites, bosses, empires. The great master expresses that this release must be a release of conscience [58, 140, 141]. Making the student a critical person who thinks about their reality and more importantly, they can transform reality [137, 138]. And I add another concept to this definition by Freire: in addition to all this, this student has to be innovative.

Within this context [142] presents an excellent article delimiting three paradigms, where Paulo Freire's vision refers to the most advanced paradigm and certainly the one with the best aspect for our civilization, the paradigm in which AI is used to empower learning while learners take agency to learn in aspect learner-as-leader. According [142] this educational paradigm reflects the theory of complexity where a synergistic collaboration between learner, instructor, information, and technology are essential and ensure a critical and innovative awareness in children and young people.

This education teaches the student to learn, and learning leads him to generate ruptures, counteract paradoxes, and invent Paulo Freire's viable novelty. The disruptions and impacts significant to the student will never come from an educator robot with clear limits. Instead, these events will come from a teacher who cries, who laughs, who feels love. A teacher that understands each human student's inner contradictions, a teacher is aware of being a person dealing with people.

I conceive this education as more similar to postgraduate guidance between the advisor and their mentee; it is private, close, and alive. It is something like "*Master Yoda*" and "*Luke*," and look how interesting it is: *Luke* had two educator robots [143]. In my opinion, educational robotics will not reduce the places for teachers in the world; it will increase because the classes of this type of education will have to be very small; therefore, I predict that in a short period, the most desired professionals by humanity will no longer be specialists in ICT or social media, will be specialist professors in paradigm three.

Considering what protects doctors, lawyers, and professors from being undermined by AI and robotic systems is human interaction based on a conscious and critical stance. It is evident that the technopositivist Cartesian view is an illusion against the possible social impacts of AI and robotics and that throughout the debate raised by the best-selling authors [38–40] the fulcrum of the argument does not address the main human characteristic that shaped humanity and characterizes every modern industrialized society: creativity.

These authors barely come close to this debate, they ignore it and go so far as to say that creativity is a parallel and secondary phenomenon and [39] it even goes so far as to claim that it is entirely explained by biochemistry. If this opinion were not ridiculous, it would be outrageous, but the worst thing is that this paradigm is taken as an undisputed axiom by the population, especially fans of these authors, although all who know biochemistry, research biochemistry, and more, research neuroscience, psychology, psychiatry, and all cognitive sciences [26, 67, 86, 87, 144], know that biochemistry does not even explain consciousness, let alone creativity.

In the book with the most beautiful title, I have ever read [145], explains that creativity is not a natural gift; therefore, at the outset, he discards linking creativity with biochemistry and conventional neurological processes. The author states that creativity is the result of attention, encouragement, overcoming the fear of making mistakes, varied and dispersed information, practice, and learning mainly from past mistakes and technical knowledge in-depth in an area of some knowledge. References [8, 45] they also make it clear that creativity is essentially a conscious phenomenon closely linked to the social phenomenon of human beings. In this context, technopositivism about AI and robotics resoundingly fails in its worldview, because in the whole core of arguments about the advances that have been made in the past and that will be made in the future, the role of creativity in the discussion is not even considered, because an AI or robot will never be able to create in the creative sense of the term.

Karnal [145] lists an ambitious and undoubtedly incomplete itinerary to stimulate creativity. However, the author's primary phrase about this itinerary is that when asked to give a more detailed itinerary, he states that if the reader needs more detail, he probably lacks: creative impulse. Moreover, here is another troubling question: What actions should education, attitudes, medicine, and considerations should advocacy (and all professions "threatened" by AI and robotics) take to enable people to have this creative impulse?

Many of the social disruptions caused by the impacts of AI and robotics may be due to humanity's misunderstanding that the advances resulting from these technologies are inflection points in our society generated mainly by the creative impulse of individuals and sometimes by collectives [68, 146], *just as it happened in the age of steam power* [8], *later on electric power* [45, 147–149], today is happening with the current world [150–152]. The rationalist-Cartesian view at the time of the steam engine stated that the human being could not support speeds greater than 80 km/h. When electric public lighting was universalized, it was believed that there would be no more robberies. Today it is preached that people will be useless or slaves in the face of AI robotics. These naive and childish prognoses quickly proved false. Reference [10] exposes the sociological perception that the way we impose a worldview on the socialization of children generates a predisposition to accept the status quo that delimits the social stratification system and justifies poverty, machismo, exclusion, racism, local and global inequality. To what extent and who is interested in painting the current unethical imbalance [29, 30, 32, 127] of modern industrialized societies as caused in the future by robotics and AI?

Perhaps to answer this question, an ideological bias is needed. I do not intend to take this direction; I choose a path with fewer paradoxes and pitfalls, something safer: ethics. Although I know that ethics reflect the ideology that supports it [153] I will direct the discussion to possible ethical limits of robotics and AI in the context of modern industrialized societies.

The word ethics comes from the word Greek ethos, which means "habit". According [154] it is a part of the philosophy that investigates the principles that motivate, discipline, guide, or distort human behavior according to social reality norms, values, and prescriptions. Are the set of rules and precepts of the moral order of a society. Reference [127] defines it as the rules of conduct derived from the sense of belonging. Reference [155] delimit that there is a subtle difference in "right or wrong" considerations between ethics and morals: ethics is associated with our relationship with others, our public dimension, moral refers to our conscience, our relationship with ourselves.

In this context, an example of ethical issues regarding the use of robotics and AI and the possible imbalance of the human/robot relationship is given by Olszewska et al. [156] express the concern that elderly care technologies could become a threat to privacy, daily interpersonal contact, or citizens' control over their own lives, making the elderly feel treated as objects rather than humans. One finding of these authors is that few standards tackle the interaction with humans and that ontologies are an efficient approach to disambiguate knowledge used among groups of humans, robots and other artificial systems that share the same conceptualization.

Veruggio and Operto [155] quote Nobel prize winner Joseph Rotblat that said *"Thinking computers, robots endowed with artificial intelligence and which can also replicate themselves (...) this uncontrolled self replication is one of the dangers in the new technologies."* Less dramatically, they point to the need to introduce ethical rules in technological applications, especially regarding the behavior of intelligent machines.

And the starting point for establishing these rules is the definition of roboethics by Veruggio and Operto [155]: *"Roboethics is an applied ethics whose objective is to develop scientific/cultural/technical tools that can be shared by different social groups and beliefs. These tools aim to promote and encourage the development of Robotics for t'he advancement of human Society and individuals, and to help preventing its misuse against humankind."*

However Mitchell [120] makes clear how difficult it is to define these rules in the context of robotics and AI, exemplifying the controversial and ethical problem of the trolley. According to her, the progress of the moral intelligence of computers cannot be separated from the progress in other kinds of intelligence, as the real challenge is to create machines that can "understand" the situations in which they are faced; the challenge in moral issues is to recognize the cause-effect relationships and from a judgment based on common sense, establish different possible futures.

These difficulties accord [155] can get involved in dual-use technology; anthropomorphization of technological products; humanization of the human–machine relationship; technology addiction; digital divide, socio-technological gap; fair access to technological resources; the effects of technology on the global distribution of wealth and power; the environmental impact of technology.

Furthermore, the same authors define the rules to be followed to try to avoid these problems and establish the limits of robotics and AI: human dignity and human rights; equality, justice, and equity; benefit and harm; respect for cultural diversity and pluralism; nondiscrimination and nonstigmatization; autonomy and individual responsibility; informed consent; privacy and confidentiality; solidarity and cooperation; social responsibility; sharing of benefits; responsibility towards the biosphere; obligatory cost–benefit analysis; exploiting potential for public discussion.

These rules are obvious and logical in democratic societies; however, as already written and referenced in this book, human relations are not logical [131–133] além de serem regidas pelas leis do Chaos [72, 78, 92] that influence the structure of society through power relations [42, 121]. So something beyond ethics is needed to avoid dystopian scenarios with robotics and AI.

Chapter 5
Conclusion

> *Love is a mortal display of immortality.*
> *Fernando Pessoa.*

A counterpoint to these extremist unethical views focused on robotics and AI is to believe, finally, in one of the guiding lights of democratic societies: the Universal Declaration of Human Rights [57]. It must always be used as a basis for decisions regarding the influence and power that humanity is willing to allow automata systems. It is essential to leave as a cornerstone of the fact that human beings must be free, conscious, and hopeful. And his principal attributes are generosity, compassion, and love.

Thus, robotic and AI systems can never have any preponderance in these fundamentals. After all, they are the fruit of human genius; they are our youngest "children," our technological youngest, so we must have the responsibility to direct them to actions that "make us proud." We want to shed tears of satisfaction with their accomplishments that are ours too and not to cry tears of blood.

Paulo de Tarso, in his magnificent first letter to the Corinthians, in chapter 13 [104], describes love in a beautiful poem that inspires even atheists. This biblical passage has some things very representative of why we do not have to fear robotics and AI.

First, this poem describes something unique and special about human beings: love αγαπε (agape)—guided by principles. It is not loved ερως (eros)—sensual and sexual, nor love φιλια (filia)—for brother and friend or love στοργη (storge)—for children and family. This agape love is that of all the saints of all religions. It is intellectual love, the one that according to [105] has to do with the mind: it is not an emotion that spontaneously arises in our hearts, but of a principle by which we deliberate.

It is this love that Francis of Assis referred to when he said: "Where there is hatred that I take love" [106]. The exciting thing is this kind of love is not only reserved for God. It is linked to the mathematician's passion for mathematics, a physicist to physics, artist to his masterpiece, musician to his music, engineer to his machine, the physician for healing, the philosopher to wisdom, and the sociologist to society.

It is thanks and only through agape that humanity has come to robotics and AI. And it will be through agape that human society will find its way into the future.

Only the human being can feel this love, and he motivates him to continue doing science, making art and making people live. It is the source of all creativity and innovation, and it is through it that the future comes. We do not need to fear brilliant robotic systems, because if they become knowledgeable, they will understand by logical deduction that the source of everything is in humans and their societies, that without us, they have no function or objective. After all, there will be no more agape if we do not exist.

Informally they will stop when encounter the paradoxical knowledge that Paulo de Tarso expressed: "Because now we see through a mirror in an enigma, but then we will see face to face; now I know in part, but then I will know as I am also known. Now, then, faith, hope, and love remain, these three; but the largest of these is agape" [104].

And yet, despite the love, there is a real risk that society will generate chaotic and complex conjunction of situations and results in which the decisions made by robots that may converge in dystopian scenarios painted by the pessimists are involved. Scenarios caused by the misuse of AI and robotic systems, such as the panoramas painted by Francisco and Francisco Pilon [29] and Boyd and Holton [27] described by Morgan [160] and interpreted by the television series Altered Carbon [107].

But we must not forget that contrary hypotheses, as well as positive ones, are only likely, and there is no certainty in anything, only hope. The quantum mechanics, which determines the real order of the universe, explains that the observer influences the probability density of a phenomenon, as indicated by Schrodinger's cat [76, 108, 109], that is, let us roll Einstein's dices [110] and hope that Smeagol stumbles [111].

References

1. Tolkien JRR (1980) Unfinished tales of númenor and middle-earth. George Allen & Unwin, London
2. Schuelke-Leech BA (2018) A model for understanding the orders of magnitude of disruptive technologies. Technol Forecast Soc Chang 129:261–274. https://doi.org/10.1016/j.techfore.2017.09.033
3. Al-Razouki M (2016) Seven global medical technology trends
4. Prevett R (2016) 15 Disruptive technology trends to watch in 2017—disruption hub. https://disruptionhub.com/15-disruptive-technology-trends-watch-2017/. Accessed 22 Aug 2019
5. Bort J (2016) 9 Tech trends from 2017 that will make billions—Business Insider. https://www.businessinsider.com/9-tech-trends-2017-billions-2016-10. Accessed 22 Aug 2019
6. Cag D (2015) 11 Awesome disruptive technology examples 2019. https://richtopia.com/emerging-technologies/11-disruptive-technology-examples. Accessed 22 Aug 2019
7. Weller C (2016) 11 Tech trends that will define 2017—Business Insider. https://www.businessinsider.com/tech-trends-that-will-define-2017-2016-12. Accessed 22 Aug 2019
8. Pinto ÁV (2005) Technology concept, vol 1. Contraponto, Rio de Janeiro
9. Camargo O (2020) Sociedade. Conceito de sociedade—Brasil Escola. In: Brasil Escola. https://brasilescola.uol.com.br/sociologia/sociedade-1.htm. Accessed 27 Feb 2020
10. Giddens A (2012) Sociology, 6th edn. Penso, Porto Alegre
11. Siciliano B, Khatib O (2008) Springer handbook of robotics. Springer, Berlin, Heidelberg
12. Hertzberg J, Chatila R (2008) AI Reasoning methods for robotics. Springer handbook of robotics. Springer, Berlin, Heidelberg, pp 207–223
13. Calsaverini RS (2009) Lógica Bayesiana. In: Ars Physica. https://arsphysica.wordpress.com/2009/11/13/logica-bayesiana/. Accessed 20 Apr 2018
14. Albiero D (2019) Agricultural robotics: a promising challenge. Curr Agric Res J 7:01–03. https://doi.org/10.12944/CARJ.7.1.01
15. Olivares-Alarcos A, Beßler D, Khamis A et al (2019) A review and comparison of ontology-based approaches to robot autonomy. Knowl Eng Rev 34. https://doi.org/10.1017/S0269888919000237
16. Fiorini SR, Bermejo-Alonso J, Goncalves P et al (2017) A suite of ontologies for robotics and automation [industrial activities]. IEEE Robot Autom Mag 24:8–11. https://doi.org/10.1109/MRA.2016.2645444
17. Olszewska JI, Barreto M, Bermejo-Alonso J et al (2017) Ontology for autonomous robotics. In: RO-MAN 2017—26th IEEE International symposium on robot and human interactive communication 2017-Janua, pp 189–194. https://doi.org/10.1109/ROMAN.2017.8172300
18. Langley P, Laird JE, Rogers S (2009) Cognitive architectures: research issues and challenges. Cogn Syst Res 10:141–160. https://doi.org/10.1016/J.COGSYS.2006.07.004

© The Author(s), under exclusive license to Springer Nature Switzerland AG 2022
D. Albiero, *Robots and AI: Illusions and Social Dilemmas*,
SpringerBriefs in Computational Intelligence,
https://doi.org/10.1007/978-3-030-95790-2

19. Olszewska JI (2019) Designing transparent and autonomous intelligent vision systems. In: ICAART 2019—proceedings of the 11th international conference on agents and artificial intelligence, vol 2, pp 850–856. https://doi.org/10.5220/0007585208500856
20. Calzado J, Lindsay A, Chen C et al (2018) SAMI: interactive, multi-sense robot architecture. In: INES 2018—IEEE 22nd international conference on intelligent engineering systems, proceedings, pp 000317–000322. https://doi.org/10.1109/INES.2018.8523933
21. Robertson H-J (2016) Toward a theory of negativity: teacher education and information and communications technology. J Teach Educ 54:280–296. https://doi.org/10.1177/002248710 3255499
22. Larson EJ (2021) The myth of artificial intelligence: why computers can't think the way we do. Harvard University Press, Cambridge
23. Russell S, Dewey D, Tegmark M (2016) Research priorities for robust and beneficial artificial intelligence
24. Capra F (2005) The hidden connections: a science for sustainable living. Cultrix, São Paulo
25. Phipps (2014) Evolutionaries. Cultrix, São Paulo
26. Smith W (2008) Cosmos and transcendence: breaking through the barrier of scientistic belief, 2th edn. Sophia Perennis, San Rafael
27. Boyd R, Holton RJ (2018) Technology, innovation, employment and power: does robotics and artificial intelligence really mean social transformation? J Sociol 54:331–345. https://doi.org/10.1177/1440783317726591
28. Matos P de C (2015) Literature review types. Faculdade de Ciências Agronomicas UNESP Campus de Botucatu 9
29. Francisco A, Francisco Pilon A (2016) Munich personal RePEc archive a global voice for survival: an ecosystemic approach for the environment and the quality of life a global voice for survival: an ecosystemic approach for the environment and the quality of life. University Library of Munich
30. Pilon AF (2018) A global voice for survival: an ecosystemic approach for the environment and the quality of life. SSRN Electron J. https://doi.org/10.2139/ssrn.3204460
31. Albiero D, Cajado D, Fernandes I et al (2015) Agroecological technologies for the semiarid region. UFC, Fortaleza
32. Branco SM (2014) Ecossistemic. Blucher, São Paulo
33. Albiero D, Rodrigues GB, Melo RP, Souza-Esquerdo VF de (2021) Agroecology interaction between university and community. Int J Outreach Commun Engagem 1:1–16. https://doi.org/10.20396/ijoce.v2i1.13971
34. Brynjolfsson E, McAfee A (2014) The second machine age: work, progress, and prosperity in a time of brilliant technologies. WW Norton, New York
35. Susskind RE, Susskind D (2015) The future of the professions: how technology will transform the work of human experts. Oxford University Press, Oxford
36. Spencer D (2017) Work in and beyond the second machine age: the politics of production and digital technologies. Work Employ Soc 31:142–152. https://doi.org/10.1177/095001701664 5716
37. Machado R (2015) Microphysics of power—Michel Foucault. Paz e Terra, São Paulo
38. Harari YN (2017) Homo deus: a brief history of tomorrow. HarperCollins, New York
39. Dawkins R (2006) The selfish gene, 30 th. Oxford University Press, Oxford
40. Ray K (2005) The singularity is near: when humans transcend biology. Penguin, New York
41. Ford MR (2015) Rise of the robots: technology and the threat of a jobless future. Basic books, New York
42. Morin E (2005) Ciencia com Consciencia. Bertrand Brasil, Rio de Janeiro
43. Platão BJ (2003) Mênon. Loyola, São Paulo
44. Kant I (2003) The critique of pure reason. Martin Claret, São Paulo
45. Pinto ÁV (2005) Technology concept, vol 2. Contraponto, Rio de Janeiro
46. Lao Tzu, Wilhelm R (1997) Tao-te king. Pensamento, São Paulo
47. Kaliakatsos-Papakostas MA, Floros A, Vrahatis MN (2013) Intelligent music composition. Swarm Intell Bio-Inspired Comput 239–256. https://doi.org/10.1016/B978-0-12-405163-8. 00010-7

48. Kaleagasi B (2017) A new ai can write music as well as a human composer. https://futurism.com/a-new-ai-can-write-music-as-well-as-a-human-composer. Accessed 23 Aug 2019

49. Marshall C (2018) Artificial intelligence writes a piece in the style of bach: can you tell the difference between JS bach and AI bach? Open culture. http://www.openculture.com/2018/01/artificial-intelligence-writes-a-piece-in-the-style-of-bach.html. Accessed 23 Aug 2019

50. Anthony S (2014) A new (computer) chess champion is crowned, and the continued demise of human grandmasters—extremetech. https://www.extremetech.com/extreme/196554-a-new-computer-chess-champion-is-crowned-and-the-continued-demise-of-human-grandm asters. Accessed 23 Aug 2019

51. Simon HA, Schaeffer J (1992) Chapter 1 the game of chess. Handb Game Theory Econ Appl 1:1–17. https://doi.org/10.1016/S1574-0005(05)80004-9

52. Bilalić M, McLeod P, Gobet F (2007) Does chess need intelligence?—A study with young chess players. Intelligence 35:457–470. https://doi.org/10.1016/J.INTELL.2006.09.005

53. Venkatasubramanian V, Luo Y, Zhang Z (2019) Control of complex sociotechnical systems: importance of causal models and game theory. Comput Chem Eng 123:1–11. https://doi.org/10.1016/J.COMPCHEMENG.2018.12.010

54. Synced (2018) AI's growing role in musical composition—syncedreview—medium. https://medium.com/syncedreview/ais-growing-role-in-musical-composition-ec105417899. Accessed 23 Aug 2019

55. Viana A (2010) Johann sebastian bach (1685–1750). http://albertoviana.net/classicos/compos itores/jsbach.htm. Accessed 23 Aug 2019

56. Garattoni B (2021) Six big AI myths. São Paulo

57. U.N. (2018) Universal declaration of human rights. https://www.unicef.org/brazil/pt/resour ces_10133.htm. Accessed 4 May 2018

58. Freire P (2013) Pedagogia da Esperança. Paz e Terra

59. Navlakha S, Bar-Joseph Z (2019) Algorithms in NATURE. In: Algorithms in nature. http://www.algorithmsinnature.org/. Accessed 1 Aug 2020

60. Jenkins R (2002) Churchill: a biography. Plume

61. Descartes R, Cress DA (1998) Discourse on method; and: meditations on first philosophy. Hackett Pub

62. Schroeder MR (1991) Fractals, chaos, power laws: minutes from an infinite paradise. Dover, New York

63. Gleick J (2014) Chaos: making a new science, 4th edn. Gradiva, Lisboa

64. Folloni A (2016) Introduction at complex teory. Juruá, Curitiba

65. Marconi M de A, Lakatos EM (2017) Fundamentos de metodologia científica. Atlas

66. Sheldrake Rupert (2012) The science delusion: freeing the spirit of enquiry. Coronet

67. Smith RD (1998) Social structures and chaos theory. Sociol Res Online 3:82–102. https://doi.org/10.5153/sro.113

68. Kuhn TS (2012) The structure of scientific revolutions: 50th anniversary edition, 4th edn. University of Chicago Press, Chicago

69. Commelin P (1985) Greco-roman mythology. Ediouro, Rio de Janeiro

70. Pérez-Villa A, Pietrucci F, Saitta AM (2018) Prebiotic chemistry and origins of life research with atomistic computer simulations. Phys Life Rev. https://doi.org/10.1016/J.PLREV.2018.09.004

71. Nussenzveig HM, Palis Júnior J (1999) Complexidade e caos. Editora UFRJ/COPEA, Rio de Janeiro

72. Prigogine I (2002) The chaos laws. UNESP, São Paulo

73. Leonel ED (2019) Invariância de Escala em Sistemas Dinâmicos Não Lineares—Editora Blucher. Blucher, São Paulo

74. Prado CPC do, Fiedler-Ferrara Nelson (1994) Caos. Blucher, São Paulo

75. Sengupta A (2006) Chaos, nonlinearity, complexity. Springer, Berlin Heidelberg

76. Alligood KT, Sauer T, Yorke JA (1997) Chaos: an introduction to dynamical systems. Springer

77. Figueiredo Camargo R, Oliveira EC de (2015) Cálculo Fracionário. Livraria da Física, São Paulo

78. Prigogine I (1980) From being to becoming: time and complexity in the physical sciences. W H Freeman & Co, New York
79. Faber J, Koppelaar H (1994) Chaos theory and social science: a methodological analysis. Qual Quant 28:421–433. https://doi.org/10.1007/BF01097019
80. Prigogine I (1982) Only an illusion. In: The tanner lectures on human values. University of Utah, Jawaharlal Nehru University, p 30
81. NHGRI (2019) The human genome project-NHGRI. https://www.genome.gov/human-gen ome-project. Accessed 26 Aug 2019
82. International-Human-Genome Sequencing-Consortium (2001) Initial sequencing and analysis of the human genome. Nature 409:860–921. https://doi.org/10.1038/35057062
83. Wirth T, Parker N, Ylä-Herttuala S (2013) History of gene therapy. Gene 525:162–169. https://doi.org/10.1016/j.gene.2013.03.137
84. Lancet-Oncology (2003) Acceleration of cure or optimisation of care? Lancet Oncol 4:261. https://doi.org/10.1016/S1470-2045(03)01059-3
85. Francis RC (2011) Epigenetics: the ultimate mystery of inheritance. Norton, New York, W.W
86. Damasio A (2000) The feeling of what happens: body and emotion in the making of consciousness. Vintage 2000, London
87. Damasio A (2018) The strange order of things: Amazon.co.uk: Antonio Damasio: 9780307908759: Books. Penguin, New York
88. Salles JF de, Haase VG, Mally-Diniz LF (2016) Development neuropyschology. Artmed, Porto Alegre
89. Brandão ML (2012) Psichophisiology. Síntesis, São Paulo
90. Mithen SJ (1998) The prehistory of the mind: a search for the origins of art, religion, and science. Phoenix, London
91. Eisberg RM, Resnick R (1985) Quantum physics of atoms, molecules, solids, nuclei, and particles. Wiley, New York
92. Mayntz R (1997) The impact of chaos on science and society: 16. Chaos in society: reflections on the impact of chaos theory on sociology. In: Grebogi C, Yorke JA (eds) The impact of chaos on science and society. United Nations University Press, New York, p 416
93. Little D (2008) Chaos and coordination in social life. Understanding Society. https://undsoc.org/2008/03/09/chaos-and-coordination-in-social-life/. Accessed 24 Jul 2020
94. LeBaron B (2020) Has chaos theory found any useful application in the social sciences? In: Scientific American—ask the experts. http://people.brandeis.edu/~blebaron/ge/chaos.html. Accessed 24 Jul 2020
95. Souza J (2018) The middle class in the mirror: its history, its dreams and illusions, its reality. Estação Brasil, Rio de Janeiro
96. Xavier RS, Galvão CB, Rodrigues RL et al (2021) Mechanical properties of lettuce (*Lactuca sativa* L.) for horticultural machinery design. Scientia Agricola 79:2022. https://doi.org/10.1590/1678-992X-2020-0249
97. Gregg J (2019) Decisions structures. In: Lawrence University. http://www2.lawrence.edu/fast/GREGGJ/CMSC150/013Decisions.html. Accessed 27 Aug 2019
98. Slater BH (2019) Logical Paradoxes. Internet Encyclopedia of Philosophy
99. Saka P (2019) Pascal's wager about god. Internet Encyclopedia of Philosophy
100. Pearce KL (2019) Omnipotence. Internet Encyclopedia of Philosophy
101. Horn RE (2019) Traps of formal logic and dialectics. Stanform University. https://web.stanford.edu/~rhorn/a/topic/phil/artclTrapsOfFormalLogic.html. Accessed 27 Aug 2019
102. Whitehead AN, Russell B (1963) Principia mathematica, vol I. Cambridge
103. Nagel E, Newman J, Hofstadter DR (2008) Godel's proof. New York University Press, New York
104. Debroy B, Debroy D (2006) The holy vedas: rig veda, yajur veda, sama veda, atharva veda, 9th. B.R. Pub. Corp. A division of D.K. Publishers Distributors (P) Ltd., Delhi
105. Wylen V, Sonntag R, Borgnakke C (1995) Fundamentals of thermodynamics, 4th edn. Blucher, São Paulo
106. Cataldo Neto A (2013) Psychiatry for medical students. EDIPUCRS, Porto Alegre

107. Carroll DW (1988) Programming in turbo pascal. McGraw Hill, São Paulo
108. Medina M, Fertig C (2006) Algorithms and programming. Novatec, p 384
109. Aigner M, Ziegler GM (2010) Proofs from the book. Springer, New York
110. Heath TL, Densmore D (2002) Euclid's elements: all thirteen books complete in one volume: the Thomas L. Heath translation. Green Lion Press, Ann Arbor
111. Asimov I (2018) Livros de Isaac Asimov em português (PDF)—Pensamentos Nómadas. http:// pensamentosnomadas.org/livros-de-isaac-asimov-em-portugues-36111. Accessed 4 May 2018
112. Future of Life Intitute (2015) Open Letter on autonomous weapons—Future of Life Institute. https://futureoflife.org/open-letter-autonomous-weapons/?cn-reloaded=1. Accessed 27 Aug 2019
113. Watch HR (2021) USA and the Russia don't want killer robots to be controlled by humans. https://br.financas.yahoo.com/news/rússia-e-eua-não-querem-213236215.html. Accessed 6 Sep 2021
114. Wikipedia (2020) Google. https://en.wikipedia.org/wiki/Google. Accessed 6 Mar 2020
115. Wall LD (2018) Some financial regulatory implications of artificial intelligence. J Econ Bus 100:55–63. https://doi.org/10.1016/J.JECONBUS.2018.05.003
116. Wikipedia (2020) 2001: a space odyssey (film)—Wikipedia. https://en.wikipedia.org/wiki/ 2001:_A_Space_Odyssey_(film). Accessed 6 Mar 2020
117. Nelson BJ, Dong L (2010) Nanorobotics 46. Nanorobotics. In: Handbook of nanotechnology. Springer, p 1972
118. Minari G (2021) Robôs e inteligência artificial vão controlar os seres humanos, prevê ex-Google—Canaltech. In: Canaltech. https://canaltech.com.br/inteligencia-artificial/robos-e-inteligencia-artificial-vao-controlar-os-seres-humanos-preve-ex-google-197545/. Accessed 14 Oct 2021
119. Rifkind H (2021) Can this man save the world from artificial intelligence? | Magazine | The Times. In: The Times. https://www.thetimes.co.uk/article/can-this-man-save-the-world-from-artificial-intelligence-329dd6zvd. Accessed 14 Oct 2021
120. Mitchell M (2020) Artificial intelligence: a guide for thinking humans. Picador, New York
121. von Weizsäcker C (1990) Ordnung und Chaos in der Wirtschaft. In: Gerok W (ed) Ordnung und Chaos in der unbelebten und belebten Natur. Wissenschaftliche Verlagsgesellschaft, Stuttgart, p 441
122. Kurtser P, Edan Y (2020) Planning the sequence of tasks for harvesting robots. Robot Auton Syst 131:103591. https://doi.org/10.1016/j.robot.2020.103591
123. IBM (2021) IBM Watson Health. https://www.ibm.com/br-pt/watson-health. Accessed 2 Sep 2021
124. IBM (2019) Watson education. IBM. https://www.ibm.com/watson/education. Accessed 26 Aug 2019
125. IBM (2019) Artificial intelligence in medicine. Machine learning. IBM. https://www.ibm. com/watson-health/learn/artificial-intelligence-medicine. Accessed 26 Aug 2019
126. Friedman LF (2014) IBM's Watson may soon be the best doctor in the world—business insider. https://www.businessinsider.com/ibms-watson-may-soon-be-the-best-doctor-in-the-world-2014-4. Accessed 26 Aug 2019
127. Fritjof C, Luisi PL (2016) The systems view of life: a unifying vision. Cambridge University Press, Cambridge
128. News U 2021's 100 Best Jobs in America. Best Jobs Rankings. US News Careers. https:// money.usnews.com/careers/best-jobs/rankings/the-100-best-jobs. Accessed 2 Sep 2021
129. Capra F (2004) The hidden connections: a science for sustainable living. Anchor Books
130. Skare T (2012) The sense of medicine. Revista Iátrico 31:2
131. Hoffman AJ (2020) The evolution of conflict, compassion and the social contract: a philosophical approach to human engagement. Aggress Violent Beh 55:101504. https://doi.org/10. 1016/j.avb.2020.101504
132. Ogawa A, Yamazaki Y, Ueno K et al (2007) Neural correlates of human cognitive bias resulting in illogical inference. Neurosci Res 58:S61. https://doi.org/10.1016/J.NEURES.2007.06.358

133. Georgiou N, Delfabbro P, Balzan R (2020) COVID-19-related conspiracy beliefs and their relationship with perceived stress and pre-existing conspiracy beliefs. Pers Individ Differ 166:110201. https://doi.org/10.1016/J.PAID.2020.110201

134. Dhir A, Kaur P, Rajala R (2018) Why do young people tag photos on social networking sites? Explaining user intentions. Int J Inf Manage 38:117–127. https://doi.org/10.1016/J.IJINFO MGT.2017.07.004

135. Guan C, Mou J, Jiang Z (2020) Artificial intelligence innovation in education: a twenty-year data-driven historical analysis. Int J Innov Stud 4:134–147. https://doi.org/10.1016/J.IJIS. 2020.09.001

136. Moraes SA de, Teruya TK (2007) Paulo freire e formação do professor na sociedade tecnológiCA. In: Anais do Simpósio Acadêmico 2007

137. Alencar AF de (2005) O Pensamento De Paulo Freire Sobre a Tecnologia: Traçando Novas Perspectivas. In: V Colóquio Internacional Paulo Freire. pp 1–13

138. Freire P (1997) Professora sim, tia não. Olho dágua

139. Freire P (1984) A máquina está a serviço de quem? BITS 1

140. Freire P (2014) Pedagogia da Autonomia. Paz e Terra

141. Freire P (1987) Pedagoia Do Oprimido. Paz e Terra

142. Artificial intelligence in education: the three paradigms. Elsevier enhanced reader. https://rea der.elsevier.com/reader/sd/pii/S2666920X2100014X?token=51BF175D41621F49B47EC BE283C7210C466EA90C50E228B5CFE40BE35DFC32A09439A27C8411F4EE625157E EE8A5B756&originRegion=us-east-1&originCreation=20210906112750. Accessed 5 Sep 2021

143. Disney (2019) StarWars.com. The official star wars website. https://www.starwars.com/. Accessed 26 Aug 2019

144. Foumani M, Razeghi A, Smith-Miles K (2020) Stochastic optimization of two-machine flow shop robotic cells with controllable inspection times: from theory toward practice. Robot Comput-Integr Manuf 61:101822. https://doi.org/10.1016/j.rcim.2019.101822

145. Karnal L (2021) The courage of hope. Planeta, São Paulo

146. Popper KR (2014) The logic of scientific discovery. Martino Fine, New York

147. Vogt HH, Albiero D, Schmuelling B (2018) Electric tractor propelled by renewable energy for small-scale family farming. In: 2018 13th International conference on ecological vehicles and renewable energies, EVER 2018. Institute of Electrical and Electronics Engineers Inc., pp 1–4

148. Melo RR, Antunes FLM, Daher S et al (2019) Conception of an electric propulsion system for a 9 kW electric tractor suitable for family farming. IET Electr Power Appl 13. https://doi. org/10.1049/iet-epa.2019.0353

149. Vogt HH, de Melo RR, Daher S et al (2021) Electric tractor system for family farming: increased autonomy and economic feasibility for an energy transition. J Energy Storage 40:102744. https://doi.org/10.1016/J.EST.2021.102744

150. Albiero D, Paulo RLD, Junior JCF et al (2020) Agriculture 4.0: a terminological introduction. Revista Ciencia Agronomica 51. https://doi.org/10.5935/1806-6690.20200083

151. Fernandes HR, Polania ECM, Garcia AP et al (2021) Agricultural unmanned ground vehicles: a review from the stability point of view. Revista Ciência Agronômica 51:2020. https://doi. org/10.5935/1806-6690.20200092

152. Megeto GAS, Silva AG da, Bulgarelli RF et al (2021) Artificial intelligence applications in the agriculture 4.0. Revista Ciência Agronômica 51:2020. https://doi.org/10.5935/1806-6690. 20200084

153. Viana N (2015) Hegemony and cultural struggle. Sociologia em rede 5:15

154. Oxford-Languages, Google (2021) What is ethic? In: Google. https://www.google.com/sea rch?q=o+que+é+ética&source=hp&ei=WEQ3YeymJ8jc1sQP3r61gA0&iflsig=ALs-wAM AAAAAYTdSaLNPvCd6UNnxPJABNps2cMtMFDUY&oq=o+que+é+ética&gs_lcp=Cgd nd3Mtd2l6EAMyBAgAEEMyBQgAEIAEMgUIABCABDIFCAAQgAQyBQgAEIAE MgUIABCABDIFCAAQgAQyBQgAEIAEMgUIABCA. Accessed 6 Sep 2021

155. Veruggio G, Operto F (2008) Roboethics: social and ethical implications of robotics. In: Springer handbook of robotics. pp 1499–1524
156. Olszewska JI, Houghtaling M, Goncalves PJS et al (2020) Robotic standard development life cycle in action. J Intell Robot Syst Theory Appl 98:119–131. https://doi.org/10.1007/s10846-019-01107-w
157. Hahn S, Mitch C (2017) St. Paul's letters to the corinthians. Ecclesiae, São Paulo
158. Shilyk (2018) 1 Coríntios 13:1–3—"Amor" ou "caridade"?. Shilyk. https://shilyk.wordpress.com/2016/10/18/1-corintios-131-3-amor-ou-caridade/. Accessed 4 May 2018
159. Chesterton GK (2014) São Francisco de Assis. Ecclesiae, São Paulo
160. Morgan RK (2003) Altered carbon (Takeshi Kovacs Novels). Del Rey
161. Netflix (2018) Altered carbon. https://www.netflix.com/br/title/80097140. Accessed 27 Aug 2019
162. Schrödinger E (1980) The present situation in quantum mechanics: a translation of Schrodinger's cat paradox paper. In: Proceedings of the American philosophical society proceedings of the American philosophical society 124:323–338
163. David B (1999) Causality and chance in modern physics. University of Pennslyvania Press, Philadelphia
164. Ohanian HC (2009) Einstein's mistakes: the human failings of genius. W.W. Norton & Co Inc., New York
165. Tolkien JRR (2005) The lord of the rings. Mariner Books, New York